**Reliability Evaluation of
Dynamic Systems Excited
in Time Domain**

Reliability Evaluation of Dynamic Systems Excited in Time Domain

Alternative to Random Vibration and Simulation

Achintya Haldar
University of Arizona, Tucson, Arizona, USA

Hamoon Azizsoltani
North Carolina State University, Raleigh, North Carolina, USA

J. Ramon Gaxiola-Camacho
Autonomous University of Sinaloa, Culiacan, Mexico

Sayyed Mohsen Vazirizade
Vanderbilt University, Nashville, Tennessee, USA

Jungwon Huh
Chonnam National University, Gwangju, Korea

Registered Office
John Wiley & Sons, Inc., 111 River Street, Hoboken, NJ 07030, USA

For details of our global editorial offices, customer services, and more information about Wiley products visit us at www.wiley.com.

Wiley also publishes its books in a variety of electronic formats and by print-on-demand. Some content that appears in standard print versions of this book may not be available in other formats.

Library of Congress Cataloging-in-Publication Data
Names: Haldar, Achintya, author.
Title: Reliability evaluation of dynamic systems excited in time domain : alternative to random vibration and simulation / Achintya Haldar, Hamoon Azizsoltani, J. Ramon Gaxiola-Camacho, Sayyed Mohsen Vazirizade, Jungwon Huh.
Description: Hoboken, NJ, USA : Wiley, [2023]
Identifiers: LCCN 2022047650 (print) | LCCN 2022047651 (ebook) | ISBN 9781119901648 (hardback) | ISBN 9781119901662 (adobe pdf) | ISBN 9781119901655 (epub)
Subjects: LCSH: Reliability (Engineering)–Mathematics. | Dynamics–Mathematics. | Vibration–Mathematical models.
Classification: LCC TA169 .H3523 2023 (print) | LCC TA169 (ebook) | DDC 620/.004520151–dc23/eng/20221205
LC record available at https://lccn.loc.gov/2022047650
LC ebook record available at https://lccn.loc.gov/2022047651

Cover image: Courtesy of Achintya Haldar
Cover design by Wiley

Set in 9.5/12.5pt STIXTwoText by Straive, Pondicherry, India

SKY10042586_020923

Contents

1 REDSET and Its Necessity *1*
1.1 Introductory Comments *1*
1.2 Reliability Evaluation Procedures Existed Around 2000 *2*
1.3 Improvements or Alternative to Stochastic Finite Element Method (SFEM) *2*
1.4 Other Alternatives Besides SFEM *4*
1.4.1 Random Vibration *4*
1.4.2 Alternative to Basic Monte Carlo Simulation *5*
1.4.3 Alternatives to Random Vibration Approach for Large Problems *5*
1.4.4 Physics-Based Deterministic FEM Formulation *5*
1.4.5 Multidisciplinary Activities to Study the Presence of Uncertainty in Large Engineering Systems *6*
1.4.6 Laboratory Testing *7*
1.5 Justification of a Novel Risk Estimation Concept REDSET Replacing SFEM *7*
1.6 Notes for Instructors *8*
1.7 Notes to Students *9*
Acknowledgments *9*

2 Fundamentals of Reliability Assessment *11*
2.1 Introductory Comments *11*
2.2 Set Theory *12*
2.3 Modeling of Uncertainty *14*
2.3.1 Continuous Random Variables *15*
2.3.2 Discrete Random Variables *16*
2.3.3 Probability Distribution of a Random Variable *16*
2.3.4 Modeling of Uncertainty for Multiple Random Variables *17*
2.4 Commonly Used Probability Distributions *19*
2.4.1 Commonly Used Continuous and Discrete Random Variables *19*

2.4.2   Combination of Discrete and Continuous Random Variables   *20*

2.5   Extreme Value Distributions   *20*

2.6   Other Useful Distributions   *21*

2.7   Risk-Based Engineering Design Concept   *21*

2.8   Evolution of Reliability Estimation Methods   *25*

2.8.1   First-Order Second-Moment Method   *25*

2.8.2   Advanced First-Order Reliability Method (AFOSM)   *26*

2.8.3   Hasofer-Lind Method   *26*

2.9   AFOSM for Non-Normal Variables   *31*

2.9.1   Two-Parameter Equivalent Normal Transformation   *31*

2.9.2   Three-Parameter Equivalent Normal Transformation   *33*

2.10   Reliability Analysis with Correlated Random Variables   *33*

2.11   First-Order Reliability Method (FORM)   *35*

2.11.1   FORM Method 1   *35*

2.11.2   Correlated Non-Normal Variables   *37*

2.12   Probabilistic Sensitivity Indices   *39*

2.13   FORM Method 2   *40*

2.14   System Reliability Evaluation   *40*

2.15   Fundamentals of Monte Carlo Simulation Technique   *41*

2.15.1   Steps in Numerical Experimentations Using Simulation   *42*

2.15.2   Extracting Probabilistic Information from *N* Data Points   *43*

2.15.3   Accuracy and Efficiency of Simulation   *43*

2.16   Concluding Remarks   *44*

3   **Implicit Performance or Limit State Functions**   *47*

3.1   Introductory Comments   *47*

3.2   Implicit Limit State Functions – Alternatives   *48*

3.3   Response Surface Method   *49*

3.4   Limitations of Using the Original RSM Concept for the Structural Reliability Estimation   *50*

3.5   Generation of Improved Response Surfaces   *51*

3.5.1   Polynomial Representation of an Improved Response Surface   *52*

3.6   Experimental Region, Coded Variables, and Center Point   *54*

3.6.1   Experimental Region and Coded Variables   *54*

3.6.2   Experimental Design   *55*

3.6.3   Saturated Design   *56*

3.6.4   Central Composite Design   *56*

3.7   Analysis of Variance   *56*

3.8   Experimental Design for Second-Order Polynomial   *58*

3.8.1   Experimental Design – Model 1: SD with Second-Order Polynomial without Cross Terms   *58*

3.8.2   Experimental Design – Model 2: SD with Second-Order Polynomial
          with Cross Terms *59*
3.8.3   Experimental Design – Model 3: CCD with Second-Order Polynomial
          with Cross Terms *61*
3.9   Comparisons of the Three Basic Factorial Designs *61*
3.10   Experimental Designs for Nonlinear Dynamic Problems Excited in the
          Time Domain *64*
3.11   Selection of the Most Appropriate Experimental Design *64*
3.12   Selection of Center Point *65*
3.13   Generation of Limit State Functions for Routine Design *66*
3.13.1   Serviceability Limit State *66*
3.13.2   Strength Limit State Functions *67*
3.13.3   Interaction Equations for the Strength Limit State Functions *67*
3.13.4   Dynamic Effect in Interaction Equations *68*
3.14   Concluding Remarks *69*

**4   Uncertainty Quantification of Dynamic Loadings Applied in the Time
          Domain** *71*
4.1   Introductory Comments *71*
4.2   Uncertainty Quantification in Seismic Loadings Applied in the Time
          Domain *73*
4.2.1   Background Information *74*
4.3   Selection of a Suite of Acceleration Time Histories Using PEER
          Database – Alternative 1 *75*
4.3.1   Earthquake Time History Selection Methodology *78*
4.4   Demonstration of the Selection of a Suite of Ground Motion Time
          Histories – Alternative 1 *79*
4.5   Simulated Ground Motions Using the Broadband Platform
          (BBP) – Alternative 2 *84*
4.5.1   Broadband Platform Developed by SCEC *84*
4.6   Demonstration of Selection and Validation of a Suite of Ground Motion
          Time Histories Using BPP *86*
4.7   Applications of BBP in Selecting Multiple Earthquake Acceleration Time
          Histories *88*
4.8   Summary of Generating Multiple Earthquake Time Histories
          Using BPP *91*
4.9   Uncertainty Quantification of Wind-Induced Wave Loadings Applied in
          the Time Domain *91*
4.9.1   Introductory Comments *91*
4.9.2   Fundamentals of Wave Loading *94*
4.9.3   Morison Equation *95*
4.10   Modeling of Wave Loading *96*

4.10.1	Wave Modeling Using the New Wave Theory	*97*
4.10.2	Wheeler Stretching Effect	*98*
4.10.3	Three-Dimensional Directionality	*98*
4.10.4	Summary of Deterministic Modeling of Wave Loading	*100*
4.11	Uncertainty Quantifications in Wave Loading Applied in the Time Domain	*100*
4.11.1	Uncertainty Quantification in Wave Loading – Three-Dimensional Constrained New Wave (3D CNW) Concept	*100*
4.11.2	Three-Dimensional Constrained New Wave (3D CNW) Concept	*102*
4.11.3	Uncertainty in the Wave Height Estimation	*104*
4.11.4	Uncertainty Quantification of Wave Loading	*105*
4.11.5	Quantification of Uncertainty in Wave Loading	*106*
4.12	Wave and Seismic Loadings – Comparisons	*107*
4.13	Concluding Remarks	*108*
5	**Reliability Assessment of Dynamic Systems Excited in Time Domain – REDSET**	*111*
5.1	Introductory Comments	*111*
5.2	A Novel Reliability Estimation Concept – REDSET	*113*
5.2.1	Integration of Finite Element Method, Improved Response Surface Method, and FORM	*113*
5.2.2	Increase Efficiency in Generating an IRS	*114*
5.2.3	Optimum Number of NDFEA Required for the Generation of an IRS	*115*
5.2.4	Reduction of Random Variables	*115*
5.3	Advanced Sampling Design Schemes	*116*
5.4	Advanced Factorial Design Schemes	*116*
5.5	Modified Advanced Factorial Design Schemes	*119*
5.5.1	Modified Advanced Factorial Design Scheme 2 (MS2)	*119*
5.5.2	Modified Advanced Factorial Design Scheme 3	*121*
5.6	Optimum Number of TNDFEA Required to Implement REDSET	*122*
5.7	Improve Accuracy of Scheme MS3 Further – Alternative to the Regression Analysis	*122*
5.7.1	Moving Least Squares Method	*122*
5.7.2	Concept of Moving Least Squares Method	*123*
5.7.3	Improve Efficiency Further to the Moving Least Squares Method	*124*
5.8	Generation of an IRS Using Kriging Method	*126*
5.8.1	Simple Kriging	*127*
5.8.2	Ordinary Kriging	*128*
5.8.3	Universal Kriging	*128*
5.8.4	Variogram Function	*131*
5.8.5	Scheme S3 with Universal Kriging Method	*133*

5.8.6   Scheme MS3 with Modified Universal Kriging Method   *133*
5.9   Comparisons of All Proposed Schemes   *133*
5.10   Development of Reliability Evaluation of Dynamical Engineering
         Systems Excited in Time Domain (REDSET)   *135*
5.10.1   Required Steps in the Implementation of REDSET   *136*
5.11   Concluding Remarks   *138*

6   **Verification of REDET for Earthquake Loading Applied in the Time
         Domain**   *139*
6.1   Introductory Comments   *139*
6.2   Verification – Example 1: 3-Story Steel Moment Frame with W24
         Columns   *140*
6.2.1   Example 1: Accuracy Study of All 9 Schemes   *140*
6.2.2   Verification – Example 2: 3-Story Steel Moment Frame with W14
         Columns   *147*
6.3   Case Study: 13-Story Steel Moment Frame   *151*
6.4   Example 4: Site-Specific Seismic Safety Assessment of CDNES   *160*
6.4.1   Location, Soil Condition, and Structures   *161*
6.4.2   Uncertainty Quantifications   *162*
6.4.3   Uncertainty Quantifications in Resistance-Related Design
         Variables   *162*
6.4.3.1 Uncertainty Quantifications in Gravity Load-related Design
         Variables   *165*
6.4.3.2 Selection of a Suite of Site-Specific Acceleration Time Histories   *165*
6.5   Risk Evaluation of Three Structures using REDSET   *166*
6.5.1   Selection of Limit State Functions   *166*
6.5.2   Estimations of the Underlying Risk for the Three Structures   *166*
6.6   Concluding Remarks   *172*

7   **Reliability Assessment of Jacket-Type Offshore Platforms Using REDSET
         for Wave and Seismic Loadings**   *175*
7.1   Introductory Comments   *175*
7.2   Reliability Estimation of a Typical Jacket-Type Offshore Platform   *176*
7.3   Uncertainty Quantifications of a Jacket-Type Offshore Platform   *177*
7.3.1   Uncertainty in Structures   *178*
7.3.2   Uncertainty in Wave Loadings in the Time Domain   *179*
7.4   Performance Functions   *180*
7.4.1   LSF of Total Drift at the Top of the Platform   *180*
7.4.2   Strength Performance Functions   *180*
7.5   Reliability Evaluation of JTPs   *181*
7.6   Risk Estimations of JTPs Excited by the Wave and Seismic
         Loadings – Comparison   *183*

7.7   Comparison of Results for the Wave and Earthquake Loadings   *190*
7.8   Concluding Remarks   *193*

8   Reliability Assessment of Engineering Systems Using REDSET for Seismic Excitations and Implementation of PBSD   *195*
8.1   Introductory Comments   *195*
8.2   Assumed Stress-Based Finite Element Method for Nonlinear Dynamic Problems   *196*
8.2.1   Nonlinear Deterministic Seismic Analysis of Structures   *196*
8.2.2   Seismic Analysis of Steel Structures   *196*
8.2.3   Dynamic Governing Equation and Solution Strategy   *197*
8.2.4   Flexibility of Beam-to-Column Connection Models by Satisfying Underlying Physics – Partially Restrained Connections for Steel Structures   *200*
8.2.5   Incorporation of Connection Rigidities in the FE Formulation Using Richard Four-Parameter Model   *202*
8.3   Pre- and Post-Northridge Steel Connections   *204*
8.4   Performance-Based Seismic Design   *207*
8.4.1   Background Information and Motivation   *207*
8.4.2   Professional Perception of PBSD   *208*
8.4.3   Building Codes, Recommendations, and Guidelines   *210*
8.4.4   Performance Levels   *210*
8.4.5   Target Reliability Requirements to Satisfy Different Performance Levels   *211*
8.4.6   Elements of PBSD and Their Sequences   *212*
8.4.7   Explore Suitability of REDSET in Implementing PBSD   *212*
8.5   Showcasing the Implementation of PBSD   *213*
8.5.1   Verification of REDSET – Reliability Estimation of a 2-Story Steel Frame   *214*
8.6   Implementation Potential of PBSD – 3-, 9-, and 20-Story Steel Buildings   *219*
8.6.1   Description of the Three Buildings   *219*
8.6.2   Post-Northridge PR Connections   *219*
8.6.3   Quantification of Uncertainties in Resistance-Related Variables   *219*
8.6.4   Uncertainties in Gravity Loads   *219*
8.6.5   Uncertainties in PR Beam-to-Column Connections   *220*
8.6.6   Uncertainties in Seismic Loading   *225*
8.6.7   Serviceability Performance Functions – Overall and Inter-Story Drifts   *226*
8.7   Structural Reliability Evaluations of the Three Buildings for the Performance Levels of CP, LS, and IO Using REDSET   *227*

8.7.1   Observations for the Three Performance Levels *228*

8.8   Implementation of PBSD for Different Soil Conditions *237*

8.9   Illustrative Example of Reliability Estimation for Different Soil
         Conditions *239*

8.9.1   Quantifications of Uncertainties for Resistance-Related Variables and
         Gravity Loads *240*

8.9.2   Generation of Multiple Design Earthquake Time Histories for Different
         Soil Conditions *240*

8.9.3   Implementation of PBSD for Different Soil Conditions *240*

8.10   Concluding Remarks *244*

**9   Reliability Assessment of Lead-Free Solders in Electronic Packaging
         Using REDSET for Thermomechanical Loadings *247***

9.1   Introductory Comments *247*

9.2   Background Information *249*

9.3   Deterministic Modelling of a Solder Ball *251*

9.3.1   Solder Ball Represented by Finite Elements *251*

9.3.2   Material Modeling of SAC Alloy *251*

9.3.2.1 HISS Plasticity Model *252*

9.3.2.2 Disturbed State Concept *254*

9.3.2.3 Creep Modeling *254*

9.3.2.4 Rate-Dependent Elasto-Viscoplastic Model *255*

9.3.3   Temperature-Dependent Modeling *255*

9.3.4   Constitutive Modeling Calibration *255*

9.3.5   Thermomechanical Loading Experienced by Solder Balls *256*

9.4   Uncertainty Quantification *257*

9.4.1   Uncertainty in all the Parameters in a Solder Ball *258*

9.4.2   Uncertainty Associated with Thermomechanical Loading *260*

9.5   The Limit State Function for the Reliability Estimation *260*

9.6   Reliability Assessment of Lead-Free Solders in Electronic
         Packaging *261*

9.7   Numerical Verification Using Monte Carlo Simulation *262*

9.8   Verification Using Laboratory Test Results *263*

9.9   Concluding Remarks *264*

Concluding Remarks for the Book - REDSET *266*

References *267*

Index *281*

1

REDSET and Its Necessity

1.1 Introductory Comments

A novel reliability evaluation method, denoted hereafter as REDSET (*Reliability Evaluation of Dynamic Systems Excited in Time Domain*) is proposed. As the secondary heading of the book suggests, REDSET will be an alternative to the classical random vibration and simulation techniques. REDSET is expected to address a knowledge gap that will be discussed later in this chapter to help us better meet current professional needs and/or requirements.

Pierre Simon Marquis de Laplace (1749–1827) wrote a commentary on general intelligence published as "A Philosophical Essay on Probabilities." He wrote *It is seen in this essay that the theory of probabilities is at bottom only common sense reduced to calculus; it makes us appreciate with exactitude that which exact minds feel by a sort of instinct without being able ofttimes to give a reason for it. It leaves no arbitrariness in the choice of opinions and sides to be taken; and by its use can always be determined the most advantageous choice. Thereby it supplements most happily the ignorance and weakness of the human mind.* (Laplace 1951). These timeless comments have a significant influence on every aspect of human activities. However, the concept of risk or reliability-based engineering analysis, design, and planning was initiated recently, most likely by Freudenthal (1956). As a result of comprehensive efforts by different engineering disciplines, design guidelines and codes have already been modified or are in the process of being modified. The Accreditation Board of Engineering and Technology (ABET) in the United States now requires that all undergraduate students in civil engineering demonstrate the necessary knowledge to follow or interpret the requirements outlined in design guidelines and apply them in everyday practices.

Reliability Evaluation of Dynamic Systems Excited in Time Domain: Alternative to Random Vibration and Simulation, First Edition. Achintya Haldar, Hamoon Azizsoltani, J. Ramon Gaxiola-Camacho, Sayyed Mohsen Vazirizade, and Jungwon Huh.
© 2023 John Wiley & Sons, Inc. Published 2023 by John Wiley & Sons, Inc.

1.2   Reliability Evaluation Procedures Existed Around 2000

After Freudenthal's initiative and the necessity of acquiring the required mathematical knowledge, a considerable amount of development in the reliability evaluation area was reported in the literature. The team members of the authors wrote two books earlier, published by John Wiley, titled "Probability, Reliability and Statistical Methods in Engineering Design" and "Reliability Assessment Using Stochastic Finite Element Analysis" (Haldar and Mahadevan 2000a, 2000b). The first book is now being used as a textbook all over the world. It covers basic reliability concepts for estimating the risk of structures that existed around 2000.

The reliability or risk of an engineering system is always estimated with respect to a performance requirement or limit state function (LSF). The first book presented various risk estimation methods with various levels of sophistication when LSFs are explicit and readily available that existed around 2000. However, when a structure is represented by finite elements (FEs), as in the second book, LSFs are expected to be implicit in most applications to appropriately incorporate information on different sources of nonlinearities satisfying the physics-based formulation to accurately estimate the required structural responses. The stochastic finite element method (SFEM) concept presented in the second book is capable of estimating reliability for implicit LSFs, but the required derivatives of LFSs with respect to the design variables to estimate risk are evaluated numerically. The basic SFEM concept cannot be used to estimate the risk of complex nonlinear dynamic engineering systems (CNDES) and an alternative is urgently needed. This book will fill the vacuum.

1.3   Improvements or Alternative to Stochastic Finite Element Method (SFEM)

One of the major impacts of the SFEM concept was that it helped reliability analysis of structures represented by FEs commonly used in many engineering disciplines. The basic finite element method (FEM)-based representation helps estimate the deterministic behavior of structures considering complicated geometric arrangements of elements made with different materials, realistic connections and support conditions, various sources of nonlinearity, and numerous sophisticated and complicated features a structure experiences from the initial to the failure state following different load paths in a very comprehensive way. However, the deterministic FEM formulation fails to incorporate the presence of uncertainty in the design variables and thus cannot estimate the underlying risk. The SFEM concept was proposed to capture the desirable features of the deterministic FEM formulation integrated with probabilistic or stochastic information on design variables. However, it was primarily developed for the static application of loadings. For ease of discussion, it can be denoted as SFEM-static. It was a significant improvement over other available methods around the mid-nineties.

The SFEM-static concept or its extension cannot be used to estimate the reliability of CNDES. It can be a building block for the reliability analysis of CNDES by representing structures using FEs with improved physics-based modeling techniques, incorporating recently introduced several advanced energy dissipation features, and exciting them in the time domain resulting in the SFEM-dynamic algorithm. However, it should be noted that SFEM-static was developed about three decades ago using outdated computational platforms not available at present. A considerable amount of time, money, and effort will be needed to modify it for current needs, and it may not be the best option. An improved version of SFEM-static is needed.

The risk evaluation capability of a procedure representing structures by FEs will depend on the efficiency of the deterministic FEM formulation used. The FEM representation should also be similar to the procedures used by the deterministic community in routine applications. To estimate nonlinear responses of frame structures, the displacement-based FEM is very commonly used. Almost all current commercially available computer programs are based on this concept. In this approach, shape functions are used to describe the displacements at the nodes of the elements. This will require a large number of elements to model members with large deformation expected just before failure, making it computationally very inefficient. It will be more efficient if the realistic structural behavior can be estimated using fewer elements and expressing the tangent stiffness matrix explicitly without updating it at every step of the nonlinear analysis. To address these issues, the assumed stress-based FEM can be used as an alternative, especially for frame-type structures.

The assumed stress-based FEM formulation, although mathematically more demanding, has many advantages over the displacement-based approach. In this approach, the tangent stiffness can be expressed in explicit form, the stresses of an element can be expressed and obtained directly, fewer elements are required in describing a large deformation configuration, and integration is not required to obtain the tangent stiffness. It is found to be accurate and very efficient in analyzing frame-type structures just before failure.

The probability of failure of a frame needs to be estimated using two iterative schemes, one to capture the nonlinear behavior and the other for the reliability estimation to consider uncertainty in the design variables. In developing SFEM-static, the displacement-based FEM approach was initially used. Subsequent studies indicated that it would be practically impossible to use this approach to extract reliability information for realistic large nonlinear structural systems. Although the stress-based FEM concept was at the early stage of development in the early eighties, it was incorporated in the subsequent developments of SFEM-static. The book on SFEM is essentially based on the stress-based FEM approach. The modified concept was extensively verified using available information.

Sophisticated computer programs were developed to implement the concept. They were written in the FORTRAN language available in the early eighties and no users' manual was developed. Most importantly, there is no other program using the concept readily available even today for commercial or academic research. Related subjects are not taught in most universities, and the basic concept is not fully developed to analyze complicated structural systems. The book by Haldar and Mahadevan (2000b) can be used. However, it will not satisfy the current professional needs.

In summary, it will take a considerable amount of time and effort to modify the SFEM-static programs suitable for different computer platforms widely used at present. The most attractive option will be if users can use any computer program capable of conducting nonlinear time domain dynamic analyses using any type of FEM to estimate structural responses.

1.4 Other Alternatives Besides SFEM

Several reliability-evaluation techniques for CNDES available at present will be briefly discussed next. Considering their deficiencies and current needs, a new concept is preferable.

1.4.1 Random Vibration

To study the stochastic behavior of dynamic systems, the classical random vibration approach (Lin 1967; Lin and Cai 2004) is expected to be the obvious choice. It is an extremely sophisticated but complicated mathematical concept. The basic random vibration concept and its derivatives include the First- or Second-Order Taylor Series Expansion, Neumann Expansion, Karhunen-Loeve Orthogonal Expansion, Polynomial Chaos, etc. Most of them were developed for relatively linear small systems with very few numbers of dynamic degrees of freedom (DDOFs). They are unable to consider the physics-based representation of nonlinear dynamic systems, failed to explicitly consider the statistical distribution information of system parameters, and were valid only for a small amount of randomness at the element level even when it had the potential to be significantly amplified at the system level. The dynamic loading is represented in the form of power spectral density functions and cannot be applied in the time domain as currently required for critical structures in design guidelines. Some of the novel features proposed after the Northridge earthquake of 1994 to improve dynamic response behavior of structures by making the connections more flexible cannot be incorporated (to be discussed in more detail in Chapter 8). The development of the random vibration concept was an important research topic during the latter half of the twentieth century and attracted the attention of scholars interested in applying sophisticated mathematical concepts in engineering applications. However, the overall effort had a marginal impact on the profession. A considerable knowledge gap still exists to study the stochastic behavior of CNDES. As the secondary title of the book suggests, an alternative to the classical random vibration concept is urgently needed.

1.4.2 Alternative to Basic Monte Carlo Simulation

A deterministic analysis of a realistic CNDES represented by FEs considering some of the newly developed attractive features may take several hours of computer time. To estimate the risk of low probability events using Monte Carlo simulation (MCS) is a possibility but it may require the continuous running of a computer for several years, as explained in Chapter 5. An alternative theoretical concept is necessary. As the secondary title of this book suggests, an alternative to MCS is also necessary.

1.4.3 Alternatives to Random Vibration Approach for Large Problems

Besides the random vibration approach, several uncertainty quantification methods for large computational models were proposed including the reduced order models, surrogate models, Bayesian methods, stochastic dimension reduction techniques, efficient MCS methods with numerous space reduction techniques, etc. Some of them are problem-specific, will not satisfy current needs, and will require considerable expertise to implement.

1.4.4 Physics-Based Deterministic FEM Formulation

The probability of failure estimation for CNDES implies that the risk needs to be estimated just before failure. The condition may be initiated by the failure of one or more structural elements in strength or if the system develops excessive vibration or deformation (deflection, rotation, etc.) making it not functional for which it was designed. Just before failure, the engineering systems are expected to go through several stochastic phases that are difficult to postulate. If the excitation is very irregular, like seismic or wave loading (discussed in Chapter 4) or thermomechanical loading (discussed in Chapter 9), it will add several additional layers of challenge. Most importantly, the deterministic community that makes the final approval decision requires that these loadings need to be applied appropriately in the time domain for the reliability estimation.

Additionally, the analytical models used to represent large engineering systems are very idealized. For example, the common assumption that supports and joints of a structure are rigidly connected or fully restrained (FR) is a major simplification; they are essentially partially restrained (PR) with different rigidities (discussed in more detail in Chapter 8). The uncertainties in the dynamic properties of structures are expected to be quite different when the connections are loading, unloading, and reloading in a very irregular fashion. Also, the failure of a member of highly indeterminate structures is expected to cause local failure without causing system failure. Current smart designs intentionally introduce alternate load transmission paths to avoid system failure. The recent trend of making connections more flexible in frame-type structures providing more ductility to increase the energy absorption capacity needs to be appropriately incorporated into the algorithm. An intelligent risk estimation procedure should be able to incorporate these advanced features indicating a reduction in the underlying risk.

After the Northridge earthquake of 1994, the Federal Emergency Management Agency (FEMA) advocated for using the performance-based seismic design (PBSD) concept proposed after the earthquake to reduce the economic loss by replacing the life safety criterion used in the past. It is a very sophisticated risk-based design method, but FEMA did not specify how to estimate the underlying risk. In fact, there is no method currently available to implement PBSD.

1.4.5 Multidisciplinary Activities to Study the Presence of Uncertainty in Large Engineering Systems

Several new methods were proposed to address uncertainty-related issues in large engineering applications in the recent past, including high-dimensional model representation (HDMR) and explicit design space decomposition – support vector machines (EDSD – SVM). In these studies, the general objective is to develop approximate multivariate expressions for a specific response surface (discussed in detail in Chapter 3). One such method is HDMR. It is also referred to as "decomposition methods," "univariate approximation," "bivariate approximation," "S-variate approximation," etc. HDMR is a general set of quantitative model assessment and analysis tools for capturing high-dimensional relationships between sets of input and output model variables in such a way that the component functions are ordered starting from a constant and adding higher order terms, such as first, second, etc. The concept is reasonable if the physical model is capable of capturing the behavior using the first few lower-order terms. However, it cannot be applied for physics-based time domain dynamic analysis and requires MCS to extract reliability information. EDSD can be used when responses are classified into two classes, e.g. safe and unsafe. A machine learning technique known as support vector machines (SVM) was used to construct the boundaries separating different modes of failure with a single SVM boundary and refined through adaptive sampling. It suffers similar deficiencies as HDMR.

This discussion clearly indicates that both HDMR and EDSD-SVM have numerous assumptions and limitations, and they use MCS to estimate the underlying risk. They fail to explicitly incorporate the underlying physics, sources of nonlinearities, etc., and dynamic loadings cannot be applied in the time domain.

1.4.6 Laboratory Testing

The durability or life of computer chips and solder balls was studied in the laboratory (Whitenack 2004; Sane 2007). They were subjected to thermomechanical loading caused by heating and cooling, causing significant changes in the material properties of the solder balls (discussed in more detail in Chapter 9). Conceptually, thermomechanical loading is also a time domain dynamic excitation in the presence of many sources of nonlinearity including severe material nonlinearity. Solder balls were tested in laboratories under idealistic loading conditions without explicitly addressing the uncertainty-related issues.

In general, the results obtained from laboratory testing are limited. It may not be possible to extrapolate results with slight variations in samples used or test conditions. To reduce the duration of testing, accelerated testing conditions are used, introducing another major source of uncertainty. Test results are also proprietary in nature. The reliability of solder balls cannot be analytically estimated at present. A sophisticated reliability evaluation technique should be robust enough to consider different types or forms of dynamic loadings applied in the time domain to excite any engineering systems. The proposed REDSET approach is expected to analytically extract reliability information for CNDES providing flexibility to consider issues not considered during testing.

1.5 Justification of a Novel Risk Estimation Concept REDSET Replacing SFEM

The discussions made in the previous sections identified the knowledge gap in estimating the risk of CNDES and an urgent need to replace the existing SFEM concept. Serious attempts were made to define the desirable characteristics or features and scope of a new reliability evaluation technique. Users should be able to extract reliability information by conducting a few deterministic FEM-based nonlinear time domain dynamic analyses using any computer program available to them.

Because of numerous deficiencies of the classical random vibration and the basic MCS concepts, these methods need to be replaced. This will add a new dimension to classical engineering analysis and design. More damage-tolerant structures can be designed using multiple deterministic analyses instead of one for any type of dynamic loading including seismic, wave, thermomechanical, etc. It will also provide an analytical approach instead of error-prone laboratory testing, saving an enormous amount of time and money.

With an advanced conceptual understanding of the uncertainty management areas and the availability of exceptional computational power, a transformative robust reliability evaluation approach needs to be developed to evaluate the underlying risk of CNDES for multidisciplinary applications. Boundaries of the reliability estimation techniques need to extend by integrating physics-based modeling with several advanced mathematical, computational, and statistical concepts, producing compounding beneficial effects to obtain acceptable probabilistic response characteristics/metrics/statistics. It needs to combine model reduction techniques, intelligent simulations, and innovative deterministic approaches routinely used by practicing professionals. If the fundamentals of the basic reliability evaluation procedure are sound, it should be capable to extract reliability information for different types of dynamic systems excited in the time domain.

REDSET is an innovative transformative reliability evaluation procedure that will fill the knowledge gap that currently exists in the profession.

1.6 Notes for Instructors

It will be impractical to present the reliability-based engineering analysis and design concepts from the very early stage to the present by combining the first two books (Haldar and Mahadevan 2000a, 2000b) and a novel concept REDSET presented here in one book. The table of contents will set the road map for teaching a course. To obtain the maximum benefit from this book, the readers are requested to review some of the fundamentals of the reliability analysis concepts discussed in the first book. They are expected to be familiar with the basic risk estimation concept including the first-order reliability method (FORM) for explicit LSFs and MCS. To make this book self-contained, some of the essential and fundamental topics are briefly presented in Chapter 2 of this book. Advance concepts required to implement REDSET are developed gradually and in very systematic ways. If any clarification is required, the first two books can be consulted. The first book has a solutions manual. It can be obtained free of cost from John Wiley (instructors in the United States only), the publisher of this book. To improve the readability, unless it is absolutely necessary, few references are cited in developing and discussing the basic concepts. However, an extended list of publications is provided at the end of the book.

1.7 Notes to Students

Scholars with different levels of technical background are expected to be interested in the novel risk estimation procedure REDSET for CNDES. They are encouraged to be familiar with the basic FORM approach as discussed in detail in Haldar and Mahadevan (2000a). However, to master the subject, it will be beneficial if attempts are made to solve all the problems given at the end of each chapter in the aforementioned book. To obtain the maximum benefit from this book describing REDSET, interested scholars are expected to be familiar with the basic seismic analysis procedure of structures using FEM. It will be helpful if a user first selects a small example from the book and excites the structure dynamically in the time domain to check the accuracy of the results before moving forward to consider multiple time histories. Verification of the results can also be made by using a small number of simulation cycles of MCS, say about 1000. Initially, an LSF of interest can be generated using the regression analysis. After developing the necessary skills, Kriging method can be attempted. Numerous examples are given in the book. It will be very useful if attempts are made to reproduce the results for as many problems as possible. And then extend the knowledge by estimating risk for many other problems not given in the book similar to solder balls discussed in Chapter 9.

Acknowledgments

The financial support of the US National Science Foundation (NSF) was essential in developing the REDSET concept presented in this book, and it needs to be appropriately acknowledged. The title of the latest grant was CDS&E: Theoretical Foundation and Computational Tools for Complex Nonlinear Stochastic Dynamical Engineering Systems – A New Paradigm. The work was supported by a relatively new division within NSF known as Computational and Data-Enabled Science and Engineering (CDS&E). It is an integrated division consisting of several other divisions within the NSF. The reviewers not only recommended the proposal for funding but also suggested increasing the budget considering its merit and training junior researchers to carry out the related research in the future. Some of the earlier studies on related areas were also supported by NSF including SFEM. During the study, the authors received financial assistance in the form of support of graduate students from several agencies of the government of Mexico: CONACYT, *Universidad Autónoma de Sinaloa* (UAS), and *Dirección General de Relaciones Internacionales de la Secretaria de Educación Pública* (DGRI-SEP).

Any opinions, findings, or recommendations expressed in this book are those of the authors and do not necessarily reflect the views of the sponsors.

2

Fundamentals of Reliability Assessment

2.1 Introductory Comments

Engineering design is essentially proportioning the shape and size of elements of a structural system satisfying several performance criteria under various demands. For example, a structure should be designed so that its strength or resistance is greater than the applied load effects during its lifetime. It is a classic supply and demand problem. In reality, the task can be very demanding and challenging due to the presence of numerous sources of uncertainty in the resistance and load-related design variables since they cannot be predicted with certainty during the lifetime of a structure. One can argue that it is virtually impossible to design a completely safe structure; the risk can be minimized but it cannot be zero. A nuclear power plant is expected to be much safer than ordinary buildings, bridges, and other infrastructures, but available information will indicate that they are not risk-free. It is essential then to estimate or quantify the underlying risk so that it will satisfy the "acceptable risk" criterion. The fundamentals of risk-based engineering analysis and design procedures readily available at present are briefly discussed in this chapter for ready reference. They are discussed in detail in Haldar and Mahadevan (2000a). Readers are strongly encouraged to refer to the materials in the book whenever any clarification is necessary.

Before initiating any risk-based engineering analysis and design or estimating risk for a design, it is essential that all the random variables (RVs) in the formulation are identified first and the uncertainties associated with them are appropriately quantified. Randomness in a variable indicates that it has multiple outcomes, and a specific value may occur more frequently than others. Usually, possible outcomes are within a range of measured or observed values. With the help of probability theory, all possible outcomes need to be identified for a variable for a specific application. After appropriately defining the performance requirements, it will be necessary to study how they will be satisfied considering all possibilities.

Reliability Evaluation of Dynamic Systems Excited in Time Domain: Alternative to Random Vibration and Simulation, First Edition. Achintya Haldar, Hamoon Azizsoltani, J. Ramon Gaxiola-Camacho, Sayyed Mohsen Vazirizade, and Jungwon Huh.
© 2023 John Wiley & Sons, Inc. Published 2023 by John Wiley & Sons, Inc.

This exercise will help extract reliability, safety, or risk information using probability theory. In some cases, the task can be simple but in most practical realistic cases, it can be extremely complicated. At present, several methods with various degrees of sophistication are available for the estimation of risk. However, there is a significant knowledge gap as briefly discussed in Chapter 1. Most of them will be discussed in more detail later in this book. In any case, the basic algebraic concepts routinely used for the deterministic design of engineering systems need to be replaced with probabilistic concepts. One simpler approach will be by conducting multiple deterministic analyses considering the presence of uncertainty in the formulation but the question remains is how. The transition will require some effort and time. Some of the essential knowledge of the risk-based design concept are discussed in this chapter in a review format. The background information is expected to help the readers obtain the maximum benefit in learning a novel risk estimation method using the REDSET concept proposed in this book for complex nonlinear dynamic engineering systems (CNDES).

2.2 Set Theory

The familiarity with set theory, another branch of mathematics, will help this transition. In the context of set theory, every problem must have a sample space consisting of sample points. An event of interest needs to be defined in terms of sample points. An event, generally denoted as E, contains at least one sample point. An event without a sample point is known as an empty or null set. When an event contains all the sample points in the sample space, it is known as the certain event (the sample space itself) and is generally denoted as S. Sample points not in an event E will be in the complementary event denoted as $\overline{E}$. If sample points can be counted, they belong to a discrete sample space, and it can be finite and infinite. In most cases of practical interest, sample points cannot be counted; they belong to continuous sample space. Suppose the annual rainfall for a particular city is of interest. There are upper and lower limits to the possible rainfall in a year and the design rainfall intensity is between these limits, in most cases. It need not be an integer and could have an infinite number of values between the limits. For a continuous sample space, the sample points cannot be counted, and it will be infinite.

In summary, each problem must have a sample space. A sample space consists of sample points that are mutually exclusive, as discussed later. An event needs to be defined in terms of sample points and each event must contain at least one sample point.

For the risk estimation purpose, in most cases, events need to be combined using union (OR) or intersection (AND) rules in the context of set theory. Sample points in events E_1, E_2, or both will constitute the union generally refer to as an OR combination and is denoted as $E_1 \cup E_2$. Sample points common to events E_1 and E_2, denoted as $E_1 \cap E_2$ or simply $E_1 E_2$, represent the intersection or AND combination. When the joint occurrence of events is impossible, i.e. the occurrence of one precludes the other, they are known as mutually exclusive events. Survival and failure of a structure at the same time are impossible and can be considered mutually exclusive events. When the unions of events constitute the sample space, they are known as collectively exhaustive events. When the events are mutually exclusive and their unions constitute the sample space, they are known as mutually exclusive and collectively exhaustive events. A sample space, the sample points in it, and the necessary events can be easily shown or drawn with the help of Venn diagrams. A Venn diagram provides a simple but comprehensive pictorial description of the problem and is routinely used by engineers.

Using set theory, the information on the probability of failure or survival of engineering systems can be obtained using three axioms or basic assumptions. They are Axiom 1 – the probability of an event, generally expressed as $P(E)$, will always be positive. Axiom 2 – the probability of the sample space, S or $P(S)$, will be 1.0 indicating that $0 \leq P(E) \leq 1.0$, and Axiom 3 – the union of two mutually exclusive events is equal to the summation of their individual probabilities, i.e. $P(E_1 \cup E_2) = P(E_1) + P(E_2)$. Using the second and third axioms, it can also be shown that $P(\overline{E}) = 1.0 - P(E)$. Simply stated, from the information on risk, the corresponding reliability can be easily calculated. If the events are not mutually exclusive, the probability of the union of two events can be expressed as:

$$P(E_1 \cup E_2) = P(E_1) + P(E_2) - P(E_1 E_2) \tag{2.1}$$

The probability of joint occurrence of two events in Eq. 2.1 can be expressed as:

$$P(E_1 E_2) = P(E_1 \mid E_2)P(E_2)\, P(E_2) = P(E_2 \mid E_1)\, P(E_1) \tag{2.2}$$

$P(E_1 \mid E_2)$ and $P(E_2 \mid E_1)$ are known as conditional probabilities indicating the probability of occurrence of one event given that the other has occurred. If the occurrence of one event does not depend on the occurrence of the other, it represents that the two events are statistically independent, and Eq. 2.2 becomes

$$P(E_1 E_2) = P(E_1)\, P(E_2) \tag{2.3}$$

These discussions indicate that set theory and mathematics of probability are simple and straightforward. However, they are conceptual in nature and very different from the deterministic concepts commonly used in the engineering profession. The lack of familiarity also makes it a little harder. Transforming deterministic problems into uncertainty-based formulations can be demanding and challenging. A lot of practice is required. Further discussions on set theory and mathematics of probability can be found in Chapter 2 of Haldar and Maha-devan (2000a). Each topic is explained with the help of numerous informative examples. At the end of the chapter, practice problems from several engineering disciplines are given to improve the understanding of the related topics. Since the information is readily available, it is not repeated here.

2.3 Modeling of Uncertainty

To estimate the probability of an event, the uncertainties in the RVs present in the problem need to be quantified. In general, the data of multiple occurrences of an RV need to be collected from as many sources as possible to quantify uncertainty in it. Using the available data, the statistical information on the uncertainty in terms of the mean, variance, standard deviation, coefficient of variation (COV), skewness (symmetry and un-symmetry), skewness coefficient, etc., can be evaluated. By plotting the histogram and frequency (histogram with an area of unity) diagrams, the information on the underlying distribution can be generated for both contin-uous and discrete RVs. By plotting the data on special graph papers or conducting several statistical tests [chi-square (χ^2) and Kolmogorov-Smirnov (K-S) are very common], the underlying distribution, i.e. normal, lognormal, binomial, Poisson, etc., can be established. An analytical model of the variation of the data can be developed to describe any shape or form in terms of its probability density function (PDF) for continuous RVs and probability mass function (PMF) for discrete RVs. The cumulative distribution function (CDF) corresponding to a specific PDF or PMF also can be obtained.

Throughout this book, an upper case letter, for example X, will be used to represent an RV and its realization, i.e. when $X = x$, it will be represented as a lower case letter. Analytical models used to quantify randomness in X need to be described mathematically. For a continuous RV X, its PDF and CDF are represented as $f_X(x)$ and $F_X(x)$, respectively, and they are related as:

$$f_X(x) = \frac{dF_X(x)}{dx} \tag{2.4}$$

The PDF is the first derivative of the CDF. It must be a non-negative function and theoretically, its range will be from $-\infty$ to $+\infty$. To calculate the probability of X between x_1 and x_2, the area under the PDF between the two limits needs to be cal-culated. This can be mathematically expressed as:

$$P(x_1 < X \le x_2) = \int_{x_1}^{x_2} f_X(x)\, dx \tag{2.5}$$

CDF represents $P(X \leq x)$ and indicates the area under the PDF for all possible values X less than or equal to x. Theoretically, the integration needs to be carried out from $-\infty$ to x and can be expressed as

$$P(X \leq x) = F_X(x) = \int_{-\infty}^{x} f_X(x)\, dx \tag{2.6}$$

CDF must be zero at $-\infty$ and 1.0 at $+\infty$; that is $F_X(-\infty) = 0.0$, and $F_X(+\infty) = 1.0$. The CDF is always greater than or equal to zero and is a nondecreasing function of an RV.

For a discrete RV, the PMF is represented as $p_X(x)$ instead of PDF and the integration needs to be replaced by summation to evaluate CDF. PMF consists of a series of spikes and CDF consists of step functions. Mathematically, the CDF of a discrete RV can be expressed as

$$F_X(x) = P(X \leq x) = \sum_{x_i \leq x} p_X(x_i) \tag{2.7}$$

Parameters are used to uniquely analytically define PDF and PMF. Depending upon the nature of the uncertainty, the number of required parameters can be one or more than one. In most cases, they can be estimated from the information on the mean, variance, skewness, etc., of the RV, which, in turn, depend on its PDF and PMF. General procedures used to estimate these parameters provide a better understanding of the uncertainty descriptor of an RV and are discussed next.

2.3.1 Continuous Random Variables

Denoting X as a continuous RV with PDF $f_X(x)$, its mean or expected value can be calculated as:

$$\text{Mean} = \text{Expected Value} = E(X) = \mu_X = \int_{-\infty}^{\infty} x f_X(x)\, dx \tag{2.8}$$

The corresponding variance, denoted as Var(X), is

$$\text{Variance} = \text{Var}(X) = \int_{-\infty}^{x} (x - \mu_X)^2 f_X(x)\, dx \tag{2.9}$$

The skewness can be calculated as:

$$\text{Skewness} = \int_{-\infty}^{x} (x - \mu_X)^3 f_X(x)\, dx \tag{2.10}$$

These expressions also give the physical meaning of mean, variance, and skewness. Mean is the centroidal distance of the area under the PDF from the origin; it is also known as the first moment of the area. Variance is the moment of inertia of the area under the PDF about its mean; it is also known as the second moment of the area under the PDF about its mean. Skewness is the third moment of the area under the PDF about its mean.

2.3.2 Discrete Random Variables

When X is a discrete RV with PMF $p_X(x_i)$, the following expressions can be used to calculate mean, variance, and skewness.

$$E(X) = \mu_X = \sum_{\text{all } x_i} x_i p_X(x_i) \tag{2.11}$$

$$\text{Variance} = \text{Var}(X) = \sum_{\text{all } x_i} (x_i - \mu_X)^2 p_X(x_i) \tag{2.12}$$

$$\text{Skewness} = \sum_{\text{all } x_i} (x_i - \mu_X)^3 p_X(x_i) \tag{2.13}$$

More discussions on the topic can be found in Haldar and Mahadevan (2000a).

2.3.3 Probability Distribution of a Random Variable

Modeling of uncertainties analytically in terms of mean, variance, and skewness was discussed in the previous section. It is now necessary to discuss procedures for selecting a particular distribution for an RV. In general, the distribution and the parameters to define it are estimated using available data. In practice, the choice of probability distribution may be dictated by mathematical convenience or by familiarity with a distribution. When sufficient data are available, a histogram or frequency diagram (a histogram with a unit area) can be used to determine the form of the underlying distribution. When more than one distribution may fit the data, statistical tests can be carried out. In some cases, the physical process may suggest a specific form of distribution. In the absence of sufficient data or any other information, the underlying distribution can be assumed to be uniform, triangular, trapezoidal, or any other mathematical form as discussed in Haldar and Mahadevan (2000a).

The validity of the underlying distribution can be established in several ways, including (i) drawing a frequency diagram, (ii) plotting the data on specialized probability papers, and (iii) conducting statistical tests known as goodness-of-fit tests for distribution including Chi-square (χ^2) or Kolmogorov–Smirnov (K-S) tests. In general, for engineering applications, any distribution can be described in terms of one, two, or three parameters and they can be estimated from the information on the mean, variance, and skewness of the RVs. All the parameters needed to describe a distribution can be obtained by using the method of moments (using the information on the first moment – mean, the second moment – variance, and the third moment – skewness), method of maximum likelihood (for an RV X, if $x_1, x_2, \ldots, x_n$ are the n observations or sample values, then the estimated value of the parameter is the value most likely to produce these observed values), etc.

Fortunately, statistical information on most of the commonly used RVs for engineering applications is already available in the literature. The relevant information can be collected by conducting a brief literature review. Many examples are given in this book and the information on uncertainties associated with all the RVs is appropriately summarized. These examples will also provide the required information on the uncertainties of different types of RVs.

2.3.4 Modeling of Uncertainty for Multiple Random Variables

Discussions made so far are based on the modeling of uncertainty in a single RV. However, based on the information available, it may be necessary to consider multiple RVs to formulate a particular problem. In fact, in most problems of practical interest, consideration of the presence of multiple RVs will be essential. If the rainfall at a site is considered to be an RV, then the runoff will also be an RV. The rainfall and the runoff can be modeled separately as RVs. However, it will be more appropriate to model the uncertainty jointly and may be more efficient in modeling the uncertainty information. Thus, the discussions of treating one RV at a time need to be broadened to consider multiple RVs. For ease of discussion, the modeling of uncertainty jointly for two RVs is briefly discussed next. Since continuous RVs are routinely used, the modeling of joint uncertainty for them is emphasized; relevant information for discrete RVs can be found in Haldar and Mahadevan (2000a). The extension of modeling of the uncertainty of more than two RVs is expected to be straightforward.

Considering X and Y are two continuous RVs, their joint PDF and CDF are expressed as $f_{X,Y}(x, y)$ and $F_{X,Y}(x, y)$, respectively. They are related as:

$$F_{X,Y}(x, y) = P(X \leq x, \ Y \leq y) = \int_{-\infty}^{x} \int_{-\infty}^{y} f_{X,Y}(u, v) dv \, du \tag{2.14}$$

The joint distribution must satisfy the three axioms of probability discussed earlier for a single RV. The observations made here will be applicable for the joint distribution as discussed here.

1) The PDF and PMF must be positive.
2) $F_{X,Y}(-\infty, -\infty) = 0, \quad F_{X,Y}(+\infty, +\infty) = 0.0$

   $F_{X,Y}(-\infty, y) = 0, \quad F_{X,Y}(x, +\infty) = 0.0$

   $F_{X,Y}(x, +\infty) = F_X(x), \ F_{X,Y}(+\infty, y) = F_Y(y)$

3) The CDF $F_{X,Y}(x, y)$ is always greater than or equal to zero and a nondecreasing function of X and Y.

For a single RV, its PDF or PMF can be plotted on two-dimensional graph paper. For two RVs, their joint PDF or PMF can be described by a three-dimensional plot. It will be challenging to plot joint PDF or PMF for more than two RVs.

When two RVs are statistically dependent of each other, their conditional PDF can be expressed as:

$$f_{X|Y}(x \mid y) = \frac{f_{X,Y}(x, y)}{f_Y(y)} \tag{2.15}$$

or

$$f_{Y|X}(y \mid x) = \frac{f_{X,Y}(x, y)}{f_X(x)} \tag{2.16}$$

If X and Y are statistically independent, then

$$f_{X,Y}(x, y) = f_X(x) f_Y(y) \tag{2.17}$$

The marginal PDF of X and Y can be shown to be:

$$f_X(x) = \int_{-\infty}^{\infty} f_{X,Y}(x, y)\, dy \tag{2.18}$$

or

$$f_Y(y) = \int_{-\infty}^{\infty} f_{X,Y}(x, y)\, dx \tag{2.19}$$

The estimation of any probability information using the joint distribution of multiple RVs can be challenging in some cases. Furthermore, the available information may not be sufficient to develop the joint distribution of multiple RVs. It will be more advantageous to study the information on the dependence or independence of two RVs to extract as much information as possible. This essentially leads to covariance and correlation analyses. The covariance of two RVs X and Y is denoted as $\text{Cov}(X,Y)$. It is the second moment about their respective means μ_X and μ_Y, and can be expressed as:

$$\text{Cov}(X,Y) = E[(X - \mu_X)(Y - \mu_Y)] = E(XY) - E(X)E(X) \tag{2.20}$$

$E(XY)$ can be calculated as:

$$\int_{-\infty}^{\infty} \int_{-\infty}^{\infty} xy\, f_{X,Y}(x, y)\, dx\, dy \tag{2.21}$$

If X and Y are statistically independent, then

$$E(XY) = E(X)E(X) \tag{2.22}$$

Thus, for statistically independent RVs, $\text{Cov}(X,Y)$ will be zero. The nondimensional value of covariance, known as the correlation coefficient, denoted as $\rho_{X,Y}$, can be estimated as:

$$\rho_{X,Y} = \frac{\text{Cov}(X,Y)}{\sigma_X \, \sigma_Y} \tag{2.23}$$

Values of $\rho_{X,Y}$ range between -1 and $+1$, and zero when X and Y are statistically independent or have no linear dependence between the two RVs. The information on the correlation coefficient needs to be used judiciously since it can also be zero for a nonlinear relationship. When the correlation coefficient is calculated using observed data, it is very rare to obtain its value to be exactly equal to zero, $+1$, or -1. Haldar and Mahadevan (2000a) suggested that the two RVs can be considered statistically independent if the correlation coefficient is less than ± 0.3 and perfectly correlated if it is more than ± 0.9.

For two discrete RVs, their joint CDF can be estimated from their joint PMF as:

$$F_{X,Y(x,\,y)} = \sum_{x_i \le x} \sum_{y_i \le y} p_{X,Y}(x_i,\,y_i) \tag{2.24}$$

All the equations given for continuous RVs can also be used for the discrete RVs, except that PDF needs to be replaced by PMF and integration needs to be replaced by summation.

2.4 Commonly Used Probability Distributions

Modeling of uncertainties in both continuous and discrete RVs is briefly discussed in the previous sections. Any mathematical model satisfying the properties of PDF or PMF and CDF can be used to quantify uncertainties in a RV. Fortunately, uncertainty associated with most of the variables routinely used in engineering applications has already been quantified in terms of a few commonly used distributions. Numerous computer programs are available and routinely used as "black boxes" without proper understanding of the theory behind them.

2.4.1 Commonly Used Continuous and Discrete Random Variables

For continuous RVs, normal, lognormal, and beta distributions are very common. The normal distribution is used when the uncertainty in a variable spans from minus to plus infinity. A lognormal distribution is used when the uncertainty in a variable spans from zero to plus infinity. The beta distribution is used when an RV is bounded by the lower and upper limits. Normal, lognormal, and beta distributions are described in terms of two parameters. For the beta distribution, the lower and upper limits also need to be defined. Among discrete RVs, binomial, geometric, and Poisson are very common. Only one parameter is required to describe these distributions. They are all discussed in great detail in Haldar and Mahadevan (2000a).

2.4.2   Combination of Discrete and Continuous Random Variables

For the sake of completeness, it should be noted that one discrete and other continuous distributions can be intelligently combined to extract the maximum reliability information. Consider the binomial distribution. If the probability of occurrence of an event in each trial is p and the probability of nonoccurrence is $(1-p)$, then the probability of x occurrences out of a total of n trials according to the binomial distribution can be expressed as:

$$P(X=x, n \mid p) = \binom{n}{x} p^x (1-p)^{n-x}, \quad x = 0, 1, 2, ..., n \tag{2.25}$$

where $\binom{n}{x}$ is the binomial coefficient and can be calculated as $n! / [x! (n-x)!]$.

The equation appears to be simple but its applications in real engineering problems can be very limited since a prior estimation of p must be available. In most cases, the information needs to be extracted from limited samples making it unreliable. This is a major weakness of the binomial distribution making its application very limited. Since p can only have values between 0 and 1, it can be represented by the standard beta distribution with parameters q and r. Suppose that m out of n items are found to be good giving a success rate of $p = m/n$. Haldar (1981) showed that its two parameters can be estimated as:

$$q = m + 1 \tag{2.26a}$$

and

$$r = n - m + 1 \tag{2.26b}$$

The resulting distribution is a combination of discrete and continuous RVs. It is denoted as hyperbinomial distribution by Haldar. This distribution was developed to address several major and expensive engineering deficiencies that occurred in projects during construction. It is discussed in more detail in Haldar (1981). This is a good example of how statistical information can be intelligently combined to solve real practical problems.

2.5   Extreme Value Distributions

The discussions on commonly used probability distributions will be incomplete without mentioning the extreme value distributions, particularly for engineering applications. The smallest or largest value of an RV often controls a particular design. Different sets of samples can be collected either by physical or numerical experiments. The minimum or maximum values for each sample set will give data to develop the underlying extreme value distribution (EVD). When the sample size approaches infinity, the distribution of the smallest or the largest values may asymptotically approach a mathematical distribution function in some cases if the samples are identically distributed and statistically independent. Gumbel (1958) classified three types of asymptotic EVDs for both minima and maxima, labeling them as Type I, Type II, and Type III EVDs. The Type I EVD of the largest value is also referred to as EVD in mechanical reliability engineering applications. The distribution of maxima in sample sets from a population with the normal distribution will asymptotically converge to Type I. The Type II EVD may result from sample sets from a lognormal distribution. The Weibull distribution is a Type III distribution with the smallest value.

EVDs are treated similarly with other distributions discussed earlier. They can be uniquely defined in terms of their PDF and CDF and the corresponding parameters obtained from the information on mean, variance, and COV. Once the EVD is defined, the probabilistic information can be extracted following the procedures discussed in the previous sections; similar to any other distributions. In most cases, PDF of an EVD will contain double exponential mathematical terms and can be difficult to handle analytically. The detailed mathematical aspects of EVDs can be found in the literature (Gumbel 1958; Castillo 1988; Haldar and Mahadevan 2000a).

2.6 Other Useful Distributions

In many engineering applications, there may not be enough information available to justify the use of a particular standard distribution. Based on limited information, an engineer may have an idea of the lower and upper limits of an RV but may not have enough data between these two limits to justify a specific distribution. In this situation, uniform and different forms of triangular or trapezoidal distributions can be used satisfying all the requirements of a PDF as discussed in detail in Haldar and Mahadevan (2000a). The parameters of these nonstandard distributions cannot be calculated from the samples since they are not available. However, they can be calculated using the assumed shape of the PDF. As discussed earlier, the mean is the centroidal distance and the variance is the first moment of inertia of the area about the centroidal axis.

2.7 Risk-Based Engineering Design Concept

Engineering design, in most cases, is essentially proportioning elements of a system so that it satisfies various performance criteria, safety, serviceability, and durability under various demands. The strength or resistance of an element must be greater than the demand or load effects during the service life. The resistance and the future critical loading conditions that may develop unacceptable behavior are unknown in most cases and the information on uncertainty in them needs to be appropriately incorporated at the time of the design. Obviously, predicting the future can be difficult and challenging. The basic risk-based design and challenges associated with it are conceptually shown in Figure 2.1 for a simple two-variables case. In the figure, R represents the resistance or capacity, and S represents the load effect or the demand on the system. At the design stage, both R and S are uncertain or random in nature. Both can be characterized by their means μ_R and μ_S, standard deviations σ_R and σ_S, and corresponding PDFs $f_R(r)$ and $f_S(s)$, respectively, as shown in the figure. The commonly used deterministic or nominal values of them, denoted as R_N and S_N used in the design, are also shown in the figure. It is obvious that the nominal values do not represent the underlying uncertainty in them. The nominal resistance R_N is a conservative value, perhaps several standard deviations below the mean value and the nominal load effect S_N is also a conservative value of several standard deviations above the mean value. The nominal safety factor S_N is defined as R_N/S_N and the safe design requires that it must be at least 1.0. Essentially, safety factors were used in estimating both the nominal resistance and loads without explicitly knowing them. Instead of using the safety factor for the resistance alone as in the working stress design (WSD) method, or for the loads alone as in the ultimate strength design (USD) method, it is more rational to apply safety factors to both resistance and loads. This concept was introduced in developing the first version of the risk-based design format known as the load and resistance factor design

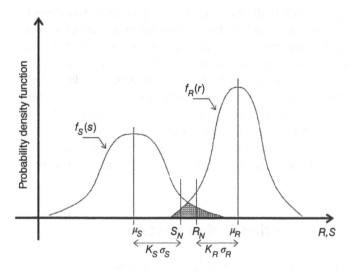

Figure 2.1 Risk-based engineering design concept.

(LRFD) concept. It was introduced in 1985 by the American Institute of Steel Construction (AISC). However, the intended conservatism introduced in the design did not explicitly incorporate information on how conservatively nominal values of R and S were selected. These shortcomings fail to convey the actual margin of safety in a design but the introduction of LRFD guidelines was a step in the right direction.

The applications of the risk-based design concept in various engineering disciplines can be considered at best as nonuniform. The following discussions are made emphasizing structural engineering applications since most of the progress had been made in this discipline. The overlapped or shaded area of the two PDF curves in Figure 2.1 provides a qualitative measure of the probability of failure. To satisfy the basic intent of the risk-based design concept explicitly incorporating information on the uncertainty in the design variables, it may not be possible to completely eliminate the overlapped area under the two PDF curves, but attempts should be made to make it as small as possible within the constraints of the economy. The objective can be achieved by considering three factors: (i) moving the relative positions of the two curves by shifting mean values μ_R and μ_S, (ii) reducing the dispersion represented by the standard deviations σ_R and σ_S, or (iii) altering the shapes of the two curves represented by the PDFs $f_R(r)$ and $f_S(s)$. Integrating all these options, the measure of risk can be expressed in terms of the probability of failure event or $P(R < S)$ as:

$$
\begin{aligned}
P(\text{failure}) = p_f &= P(R < S) \\
&= \int_0^\infty \left[\int_0^s f_R(r)\, dr \right] f_S(s)\, ds \\
&= \int_0^\infty F_R(s) f_S(s)\, ds
\end{aligned}
\tag{2.27}
$$

where $F_R(s)$ is the CDF of R evaluated at s. The equation states that when the load $S = s$, the probability of failure is $F_R(s)$, and since the load is an RV, the integration needs to be carried out for all the possible values of S represented by its PDF. This is the basic equation of the risk-based design concept. Since the CDF of R and the PDF of S may not be available in explicit forms in most cases of practical interest, the integration cannot be evaluated except for a few idealized cases as discussed by Haldar and Mahadevan (2000a). They derived expressions for the capacity reduction factor and load factors used in the prescriptive LRFD concept in terms of the underlying risk and the statistical characteristics of R and S. This information is generally unknown to the practicing design engineers.

The fundamentals of the classical reliability analysis need to be described at this stage. So far, most of the discussions made are for idealized two RVs, R and S, representing numerous resistance- and load-related RVs, respectively. R contains statistical information on several RVs related to the resistance including cross-sectional and materials properties of all structural elements in the structure under consideration, load paths to failure, etc. Similarly, S contains information on the loads or the critical load combination that may initiate the failure condition. Thus, the discussions need to be broadened to develop the basic reliability analysis concept.

The probability of failure is always estimated with respect to a specific performance criterion using relevant statistical information on load- and resistance-related design variables, known as the basic variables X_i, the functional relationship among them, and the permissible value that will satisfy the performance requirement. The performance function can be mathematically expressed as:

$$Z = g(X_1, X_2, \ldots, X_n) \tag{2.28}$$

The failure surface or the limit state of interest can be defined as $Z = 0$. It represents a boundary between the safe and unsafe regions in the design parameter space. It represents a state beyond which a structure can no longer fulfill the purpose for which it was designed. Assuming R and S are the two basic RVs, the failure surface, and safe and unsafe regions are shown in Figure 2.2. A limit state equation can be an explicit or implicit function of the basic RVs and it can be in simple linear or complicated polynomial form. It will be discussed in more detail in Chapter 3.

Considering the nature of the limit state equation, reliability evaluation methods with different levels of sophistication and complexity have been proposed. Referring to Figure 2.2, the failure can be defined as when $Z < 0$. The probability of failure, p_f, can be estimated as:

$$p_f = \int \cdots \int_{g(\ ) < 0} f_X(x_1, x_2, \ldots, x_n)\, dx_1\, dx_2, \ldots, dx_n \tag{2.29}$$

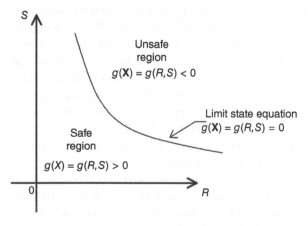

Figure 2.2 Limit state concept.

where $f_X(x_1, x_2, ..., x_n)$ is the joint PDF for all the basic RVs $X_1, X_2, ..., X_n$ and the integration is performed in the failure region $g(\ ) < 0$. If the RVs are statistically independent, then the joint PDF can be replaced by the product of the individual PDF of all the RVs.

Equation 2.29 is a more general representation of Eq. 2.27. It is known as the full distributional approach and can be considered the fundamental equation for reliability analysis. Unfortunately, the joint PDF of all the RVs present in the formulation is practically impossible to obtain. Even if this information is available, evaluating the multiple integrals can be demanding and challenging. This prompted the development of many analytical approximations of the integral that are simpler to compute for reliability estimation. These methods can be broadly divided into two groups, the first-order reliability method (FORM) and the second-order reliability method (SORM).

2.8 Evolution of Reliability Estimation Methods

It will be informative to document how the reliability estimation or assessment methods evolved with time, as briefly reviewed next.

2.8.1 First-Order Second-Moment Method

The development of FORM can be traced historically to the first-order second-moment method (FOSM). FOSM used the information on the first and second moments of the RVs for the reliability evaluation completely ignoring the distributional information. It is also known as the mean value first-order second-moment (MVFOSM) method. In MVFOSM, the performance function is linearized using first-order Taylor series approximation at the mean values of the RVs using only second-moment statistics (mean and covariance). Haldar and Mahadevan (2000a) showed that if the event of failure is defined as $R < S$ or $Z < 0$, the probability of failure will depend on the ratio of the mean value of Z, μ_Z to its standard deviation, σ_Z. The ratio is commonly known as the safety index or reliability index and is universally denoted as β. It can be calculated as:

$$\beta = \frac{\mu_Z}{\sigma_Z} \tag{2.30}$$

Using the information on β, the probability of failure can be shown to be:

$$p_f = \Phi(-\beta) = 1.0 - \Phi(\beta) \tag{2.31}$$

where $\Phi(\ )$ is the CDF of a standard normal variable.

MVFOSM was used to develop some of the earlier versions of the reliability-based design guidelines (Canadian Standard Association (CSA) 1974; Comité European du Be'ton (CEB) 1976; American Institute of Steel Construction (AISC) 1986).

There are numerous deficiencies in FOSM. It does not use the distributional information of RVs. The function Z in Eq. 2.28 is linearized at the mean values of the X_i variables. For nonlinear function, significant error may be introduced by neglecting higher order terms. Most importantly, the safety index β defined by Eq. 2.30 fails to be identical under different but mechanically equivalent formulations of the same performance function. For example, a failure event can be defined by $(R - S < 0)$ and $(R/S < 1)$, which are mechanically equivalent. The reliability indexes evaluated by FOSM for the two formulations will be different. Engineering problems can also be formulated in terms of stress or strength. However, they will not give an identical reliability index. Because of its severe limitations, alternatives to FOSM or MVFOSM were proposed, as discussed next.

2.8.2 Advanced First-Order Reliability Method (AFOSM)

Improvements to FOSM were proposed in several stages. When distributional information is incorporated in FOSM, it is known as advanced FOSM or AFOSM. More discussions are needed to clarify it further.

2.8.3 Hasofer-Lind Method

Hasofer and Lind (1974) proposed several important improvements to FOSM assuming all the variables are normally distributed. They proposed a graphical representation or interpretation of the safety index β, as defined by Eq. 2.30. Using the Hasofer-Lind (H-L) method, an RV is defined in the transformed or reduced coordinate system as:

$$X_i' = \frac{X_i - \mu_{X_i}}{\sigma_{X_i}} \tag{2.32}$$

where X_i' is a standard normal RV with zero mean and unit standard deviation. The original limit state $g(\mathbf{X}) = 0$ now can be expressed in the reduced coordinate system as $g(\mathbf{X'}) = 0$. Note that since X_i is normal, X_i' is standard normal. The safety index according to the H-L method, denoted hereafter as $\beta_{\text{H-L}}$, is defined as the minimum distance from the origin of the axes in the reduced coordinate system to the limit state surface (failure surface). It can be expressed as:

$$\beta_{\text{H-L}} = \sqrt{(\mathbf{x'^*})^t (\mathbf{x'^*})} \tag{2.33}$$

where $\mathbf{x}'^*$ is a vector containing the coordinates of the minimum distance point on the limit state surface, generally denoted as the design point or checking point, in the reduced coordinate system. It will be $\mathbf{x}^*$ in the original coordinate system. These vectors represent the values of all the RVs, that is, $X_1, X_2, ..., X_n$ at the design point corresponding to the coordinate system being used.

The H-L method gave a graphical representation of the safety index concept. The basic concept can be easily described with the help of a linear limit state function containing two variables R and S, as considered before. They need not be normal variables. Suppose the limit state equation can be expressed as:

$$Z = R - S = 0 \tag{2.34}$$

The corresponding values for R and S in the reduced coordinate system are:

$$R' = \frac{R - \mu_R}{\sigma_R} \tag{2.35}$$

and

$$S' = \frac{S - \mu_S}{\sigma_S} \tag{2.36}$$

Substituting these values into Eq. 2.34 will result in the limit state equation in the reduced coordinate system as:

$$g(\ ) = Z = \sigma_R R' - \sigma_S S' + \mu_R - \mu_S = 0 \tag{2.37}$$

The original and transformed limit state equations are plotted in Figure 2.3a, b, respectively.

The safe and unsafe regions and other relevant information are also shown in the figure. From Figure 2.3b, it can be easily observed that if the limit state line is closer to the origin in the reduced coordinate system, the failure region will be larger. The position of the limit state surface relative to the origin in the reduced coordinate system is a measure of reliability or risk. For a linear limit state equation and using trigonometry, the shortest distance between the design point and the origin can be calculated as:

$$\beta_{\text{H-L}} = \frac{\mu_R - \mu_S}{\sqrt{\sigma_R^2 + \sigma_S^2}} \tag{2.38}$$

This distance is known as the reliability or safety index; however, it was obtained in a very different way. The graphical representation of the safety index was a very significant development toward the development of modern reliability estimation procedures. Haldar and Mahadevan (2000a) showed that the reliability index would be identical to the value obtained by MVFOSM for the linear limit state equation if both R and S are normal RVs. It may not be the same for other cases.

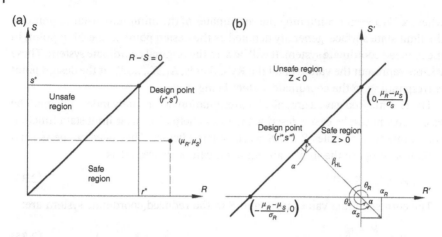

Figure 2.3 Hasofer–Lind reliability index for linear limit state function; (a) original coordinates, (b) reduced coordinates.

The H-L method needs to be extended to consider more general cases by removing other deficiencies. Suppose the vector $\mathbf{X} = (X_1, X_2, ..., X_n)$ represents several RVs in the original coordinate system and $\mathbf{X}' = \left(X_1', X_2', ..., X_n'\right)$, the same variables in the reduced coordinate system, and the limit state $g(\mathbf{X}') = 0$ is a nonlinear function as shown in Figure 2.4 for the two-variables case. They are considered to be uncorrelated at this stage. How to consider correlated RVs will be discussed in Section 2.9. Safe and unsafe regions, coordinate of the checking point $(\mathbf{x}'^*)$, and $\beta_{\text{H-L}}$ (the shortest distance from the origin to the checking point on the limit state) are shown in the figure. In this definition, the reliability index is invariant and can be estimated using Eq. 2.33. For the limit state surface where the failure region is away from the origin, it can be observed from Figure 2.4 that $\mathbf{x}'^*$ is the most probable failure point (MPFP). It represents the worst combination of the RVs and is denoted as the design point or the MPFP.

For nonlinear limit state functions, the computation of the minimum distance becomes an optimization problem. Referring to Figure 2.4, it can be mathematically expressed as:

$$\text{Minimize Distance } D = \sqrt{\mathbf{x}'^t \mathbf{x}'} \tag{2.39}$$

$$\text{Subject to constraint } g(\mathbf{x}) = g(\mathbf{x}') = 0 \tag{2.40}$$

where $\mathbf{x}'$ represents the coordinates of the checking point on the limit state equation in the reduced coordinates. Using the method of Lagrange multipliers, the minimum distance can be obtained as:

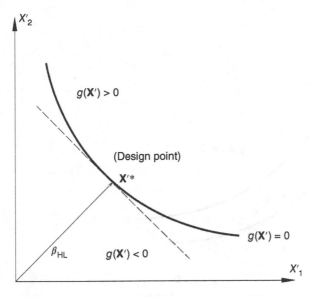

Figure 2.4 Hasofer-Lind reliability index: Nonlinear performance function.

$$\beta_{\text{H-L}} = -\frac{\sum\limits_{i=1}^{n} \mathbf{x}'^{*}\left(\dfrac{\partial g}{\partial X_i'}\right)^{*}}{\sqrt{\sum\limits_{i=1}^{n}\left(\dfrac{\partial g}{\partial X_i'}\right)^{2*}}} \tag{2.41}$$

where $\left(\partial g/\partial X_i'\right)^{*}$ is the partial derivative of the ith RV at the checking point with coordinates $\left(x_1'^{*}, x_2'^{*}, ..., x_n'^{*}\right)$. The asterisk after the derivative indicates that it is evaluated at $\left(x_1'^{*}, x_2'^{*}, ..., x_n'^{*}\right)$. The checking point in the reduced coordinates can be shown to be:

$$x_i'^{*} = -\alpha_i \beta_{\text{H-L}} \ (i = 1, 2, ..., n) \tag{2.42}$$

where

$$\alpha_i = \frac{\left(\dfrac{\partial g}{\partial X_i'}\right)^{*}}{\sqrt{\sum\limits_{i=1}^{n}\left(\dfrac{\partial g}{\partial X_i'}\right)^{2*}}} \tag{2.43}$$

$\alpha_i's$ are the direction cosines along the coordinate axes $X_i's$. In the original coordinates space and using Eq. 2.32, the coordinate for the checking point can be shown as:

$$x_i^{*} = \mu_{X_i} - \alpha_i \sigma_{X_i} \beta_{\text{H-L}} \tag{2.44}$$

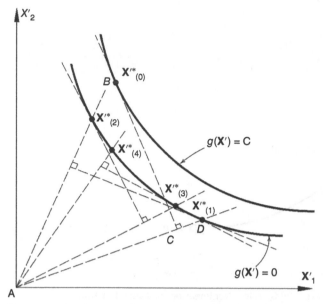

Note: A number in parentheses indicates iteration number

Figure 2.5 Iterative schemes to find the Hasofer-Lind reliability index.

An algorithm was proposed by Rackwitz (1976) following the above thought process. However, it was not ready for general applications since all the variables were still considered to be normally distributed. In any case, the basic concept is shown in Figure 2.5. The algorithm constructs a linear approximation to the limit state at every search point and finds the distance from the origin. It is iterative in nature. Point B in Figure 2.5 represents the initial checking point, usually considered to be at the mean values of all the RVs. Note that Point B is not on the limit state equation $g(X') = 0$. The tangent to the limit state at B is represented by the line BC. The distance AD will give an estimate of $\beta_{H\text{-}L}$ in the first iteration. As the iteration process continues, $\beta_{H\text{-}L}$ will converge satisfying a preassigned tolerance level.

Discussions made so far need to be summarized at this stage. The reliability indexes obtained by MVFOSM and AFOSM will be identical when the limit state equation is linear, and all the variables are normal. MVFOSM does not use any distributional information on the RVs and AFOSM is applicable when they are normal variables. The most important difference is that in MVSOFM, the checking point is at the mean values of RVs; they are not on the limit state line. For the linear limit state equation, no iteration is required; the design point will be on the limit state line. These are discussed in more detail in Haldar and Mahadevan (2000a).

2.9 AFOSM for Non-Normal Variables

The Hasofer-Lind reliability index, $\beta_{\text{H-L}}$, can be exactly related to the probability of failure using Eq. 2.31 if all the variables are statistically independent and normally distributed and the limit state surface is linear. For general applications, these requirements cannot be satisfied, and further improvements are necessary as suggested by Rackwitz and Fiessler (1978) and Chen and Lind (1983). In the context of AFOSM, the failure probability can be estimated using two types of approximations to the limit state at the checking point: first-order leading to the name FORM and second-order leading to the name SORM. MVFOSM discussed earlier is an earlier version of FORM. At present, AFOSM is known as FORM. The Hasofer-Lind method discussed earlier is an earlier version of AFOSM when all the variables are normal. The widely used FORM method of interest for this book is discussed next when the variables can be non-normal, statistically dependent, and the limit state function can be linear and nonlinear.

Conceptually, statistically independent non-normal variables can be transformed into equivalent normal variables in several ways. Almost all the RVs used in engineering applications can be uniquely described with the help of two parameters (mean and standard deviation) as emphasized in the following discussions. For three parameter distributions, methods suggested by Chen and Lind (1983) and Wu and Wirsching (1987) can be used, as discussed in Section 2.9.2.

2.9.1 Two-Parameter Equivalent Normal Transformation

A non-normal distribution cannot be transformed into a normal distribution over its entire domain. For example, the lognormal distribution is valid from zero to $+\infty$. However, a normal distribution is valid from $-\infty$ to $+\infty$. Thus, the transformation of the lognormal distribution to the normal distribution is not possible. For the ith two-parameter non-normal variable, the equivalent normal mean and standard deviation, denoted hereafter as $\mu_{X_i}^N$ and $\sigma_{X_i}^N$, can be calculated at the checking point by satisfying the following two conditions as suggested by Rackwitz and Fiessler (1976). Considering each statistically independent non-normal variable individually and equating its CDF and PDF to be the same as the equivalent normal variable at the checking point x_i^* will give the two conditions required for the estimation of $\mu_{X_i}^N$ and $\sigma_{X_i}^N$. These two conditions can be mathematically expressed as:

$$\Phi\left(\frac{x_i^* - \mu_{X_i}^N}{\sigma_{X_i}^N}\right) = F_{X_i}\left(x_i^*\right) \tag{2.45}$$

and

$$\frac{1}{\sigma_{X_i}^N} \Phi \left(\frac{x_i^* - \mu_{X_i}^N}{\sigma_{X_i}^N} \right) = f_{X_i}(x_i^*) \tag{2.46}$$

where $\Phi()$ is the CDF of the standard normal variable with an equivalent mean and standard deviation of $\mu_{X_i}^N$ and $\sigma_{X_i}^N$, respectively, and $F_{X_i}(x_i^*)$ is the CDF of the original non-normal RV at the checking point, $\varphi()$ is the PDF of the equivalent standard normal variable, and $f_{X_i}(x_i^*)$ is the PDF of the original non-normal variable at the checking point. From the above two equations, it can be shown that:

$$\mu_{X_i}^N = x_i^* - \Phi^{-1}[F_{X_i}(x_i^*)]\sigma_{X_i}^N \tag{2.47}$$

and

$$\sigma_{X_i}^N = \frac{\varphi\{\Phi^{-1}[F_{X_i}(x_i^*)]\}}{f_{X_i}(x_i^*)} \tag{2.48}$$

All the variables were defined earlier. Thus, the equivalent mean and standard deviation values can be used to implement FORM to incorporate method the distribution information of all RVs, eliminating one of the deficiencies of H-L or AFOSM.

The team members of the authors observed that the above approximation of estimating the two equivalent parameters of non-normal distributions can become more and more inaccurate if the original distribution is increasingly skewed. For a highly skewed distribution and relatively large value of x_i^*, the CDF value will be close to one and the PDF value will be close to zero. This prompted Rackwitz and Fiessler (1978) to suggest that the mean value and the probability of exceedance of the equivalent normal distribution are made equal to the median value and the probability of exceedance of the original RV, respectively, at the checking point. The two equivalent normal parameters can be calculated as:

$$\mu_{X_i}^N = F_{X_i}^{-1}(0.5) = \text{median of } X_i \tag{2.49}$$

and

$$\sigma_{X_i}^N = \frac{x_i^* - \mu_{X_i}^N}{\Phi^{-1}[F_{X_i}(x_i^*)]} \tag{2.50}$$

where $F_{X_i}^{-1}()$ is the inverse of the CDF of non-normal X_i.

Ayyub and Haldar (1984) observed that the problem might occur in many designs invalidating the original non-normal distribution. They noted that the larger x_i^* is, the smaller $\mu_{X_i}^N$ tends to be, even a negative value, not possible for some distribution destroying the validity of the distribution of X_i. They suggested a lower limit of zero for $\mu_{X_i}^N$ to obtain an accurate estimate of the reliability index β and the probability of failure p_f using the optimization algorithm of FORM. If this lower value is imposed on $\mu_{X_i}^N$, then if $\mu_{X_i}^N < 0$,

$$\mu_{X_i}^N = 0 \tag{2.51}$$

and

$$\sigma_{X_i}^N = \frac{x_i^*}{\Phi^{-1}[F_{X_i}(x_i^*)]} \tag{2.52}$$

Otherwise, use Eqs. 2.47 and 2.48.

2.9.2 Three-Parameter Equivalent Normal Transformation

Most of the distributions commonly used in engineering applications can be described with the help of two parameters. Sometimes, it may be required to deal with a distribution that needs to be described with the help of three parameters. The algorithm discussed in the previous section can be extended to consider a three-parameter approximation as suggested by Chen and Lind (1983). They introduced a third parameter A_i generally referred to as a scale factor. For each non-normal variable X_i, A_i can be calculated by imposing the condition that at the checking point, in addition to the mean and standard deviation, the slopes of the PDF must be equal for both the original and the equivalent normal distributions.

Another alternative was proposed by Wu and Wirsching (1987). They also used a scale factor A_i. It is approximately calculated as the ratio of the failure probability estimate with the original distribution to the failure probability estimate with the equivalent normal distribution by assuming a linearized limit state function and replacing the effect of all other variables with a single normal variable. Then the other two parameters $\mu_{X_i}^N$ and $\sigma_{X_i}^N$ are computed by minimizing the sum of squares of the errors in the probability estimates between the original and the equivalent normal distributions.

2.10 Reliability Analysis with Correlated Random Variables

In the reliability analysis of realistic structures, some of the RVs are expected to be correlated. The basic reliability estimation methods implicitly assume that all the variables $X_1, X_2, ..., X_n$ are uncorrelated. For correlated variables X_i's with a mean and standard deviation of μ_{X_i} and σ_{X_i}, respectively, the covariance matrix, **C**, can be expressed as:

$$[\mathbf{C}] = \begin{bmatrix} \sigma_{X_1}^2 & \text{cov}(X_1, X_2) & \cdots & \text{cov}(X_1, X_n) \\ \text{cov}(X_2, X_1) & \sigma_{X_2}^2 & \cdots & \text{cov}(X_2, X_n) \\ \vdots & \vdots & & \vdots \\ \text{cov}(X_n, X_1) & \text{cov}(X_n, X_2) & \cdots & \sigma_{X_n}^2 \end{bmatrix} \tag{2.53}$$

Defining the reduced variables by Eq. 2.32, the covariance matrix $[\mathbf{C'}]$ of the reduced variables $X_i's$ can be written as:

$$[\mathbf{C'}] = \begin{bmatrix} 1 & \rho_{X_1, X_2} & \cdots & \rho_{X_1, X_2} \\ \rho_{X_2, X_1} & 1 & \cdots & \rho_{X_2, X_n} \\ \vdots & \vdots & \vdots & \\ \rho_{X_n, X_1} & \rho_{X_n, X_2} & \cdots & 1 \end{bmatrix} \tag{2.54}$$

where $\rho_{X_i X_j}$ is the correlation coefficient of the X_i and X_j variables.

In reliability estimation, the correlated variables $X_i's$ are to be transformed into uncorrelated reduced **Y** variables and the applicable limit state equation (Eq. 2.28) needs to be expressed in terms of **Y** variables. Haldar and Mahadevan (2000a) suggested the following equation:

$$\{\mathbf{X}\} = [\sigma_{\mathbf{X}}^N] \, [\mathbf{T}] \, \{\mathbf{Y}\} + \{\mu_{\mathbf{X}}^N\} \tag{2.55}$$

where $\mu_{X_i}^N$ and $\sigma_{X_i}^N$ are the equivalent normal mean and standard deviation, respectively, of the X_i variables evaluated at the checking point, as discussed earlier, and **T** is a transformation matrix to convert the correlated reduced **X'** variables to uncorrelated reduced **Y** variables. Note that the matrix containing the equivalent normal standard deviation in Eq. 2.55 is a diagonal matrix. The transformation matrix **T** can be shown to be:

$$[\mathbf{T}] = \begin{bmatrix} \theta_1^{(1)} & \theta_1^{(2)} & \cdots & \theta_1^{(n)} \\ \theta_2^{(1)} & \theta_2^{(2)} & \cdots & \theta_2^{(n)} \\ \vdots & \vdots & & \vdots \\ \theta_n^{(1)} & \theta_n^{(2)} & \cdots & \theta_n^{(n)} \end{bmatrix} \tag{2.56}$$

$[\mathbf{T}]$ contains the eigenvectors of the correlation matrix $[\mathbf{C'}]$ represented by Eq. 2.54. $\{\boldsymbol{\theta}^{(i)}\}$ is the eigenvector of the ith mode. $\theta_1^{(i)}, \theta_2^{(i)}, ..., \theta_n^{(i)}$ are the components of the ith eigenvector. Haldar and Mahadevan (2000a) discussed all the steps in detail with the help of several examples.

For large problems, the correlated variables can also be transformed into uncorrelated variables through an orthogonal transformation of the form

$$\mathbf{Y} = \mathbf{L}^{-1} \left(\mathbf{X}'\right)^t \tag{2.57}$$

where $\mathbf{L}$ is the lower triangular matrix obtained by Cholesky factorization of the correlation matrix $[\mathbf{C}']$. If the original variables are non-normal, their correlation coefficients change on the transformation to equivalent normal variables. Der Kiureghian and Liu (1985) suggested semi-empirical formulas for fast and reasonably accurate computation of $[\mathbf{C}']$.

2.11 First-Order Reliability Method (FORM)

Based on the above discussions, it is now possible to estimate the reliability index β for all cases, instead of $\beta_{\text{H-L}}$, i.e. the performance functions can be linear or non-linear, may contain normal and non-normal RVs, and they can be uncorrelated or correlated. Using the discussions made so far, it is necessary to integrate them and formally present all the necessary required steps to implement FORM, one of the major building blocks of the novel concept REDSET that will be presented later in this book. Two optimization algorithms can be used to estimate risk or reliability using FORM. The first method, denoted hereafter as FORM Method 1, requires the solution of the limit state equation during the iterations; essentially the limit state equation is available in explicit form. FORM Method 2 does not require a solution of the limit state equation, or it can extract reliability information when the limit state function is implicit in nature. Instead, a Newton–Raphson type recursive algorithm can be used to obtain the design point. One of the major objectives of this book is to estimate reliability with implicit limit state equations. However, numerous ways the required functions can be generated making them explicit, even in approximate ways, making FORM Method 2 irreverent. If interested, FORM Method 2 is discussed in detail in Haldar and Mahadevan (2000a) in developing the SFEM concept discussed in Chapter 1. Since REDSET is expected to replace SFEM, no additional discussion on FORM Method 2 will be made here.

2.11.1 FORM Method 1

For the ease of discussion, nine steps required to implement FORM Method 1, including some improvements in the algorithm suggested by Ayyub and Haldar (1984), are presented below. The original coordinate system is used in describing these steps. All the variables are assumed to be uncorrelated.

- Step 1: Define the appropriate limit state equation. It will be discussed more in detail in Chapter 3. It is discussed briefly in Example 2.1.

- Step 2: Assume an initial value of the safety index β. For structural engineering applications, any value of β, say between 3 and 5 can be assumed; if it is chosen appropriately, the algorithm will converge in very few steps. An initial β value of 3.0 is reasonable.
- Step 3: To start the iterative process, assume the values of the checking points x_i^*, $i = 1, 2, ..., n$. In the absence of any prior information, the coordinates of the initial checking point can be assumed to be at the mean values of the RVs.
- Step 4: Compute the mean and standard deviation of the equivalent normal distribution at the checking point for those variables that are non-normal.
- Step 5: Compute partial derivatives of the limit state equation with respective to the design variables at the checking point x_i^*, i.e. $(\partial g/\partial X_i)^*$.
- Step 6: Compute the direction cosine α_i at the checking point for the X_i RV as:

$$\alpha_{X_i} = \frac{\left(\dfrac{\partial g}{\partial X_i}\right)^* \sigma_{X_i}^N}{\sqrt{\sum_{i=1}^{n} \left(\dfrac{\partial g}{\partial X_i}\, \sigma_{X_i}^N\right)^{2*}}} \tag{2.58}$$

Note that Eqs. 2.43 and 2.58 are identical. In Eq. 2.43, the direction cosines are evaluated in the reduced coordinates where the standard deviations in the reduced coordinates are unity. In Eq. 2.58, if the RVs are normal, their standard deviations can be used directly. For non-normal variables, the equivalent standard deviations at the checking point need to be used. As discussed in Step 3, for the first iteration, the coordinates of the initial checking point can be assumed to be at the mean values of the RVs.

- Step 7: Compute the new value of the checking point for the ith RV for x_i^* as:

$$x_i^* = \mu_{X_i}^N - \alpha_{X_i}\beta\, \sigma_{X_i}^N \tag{2.59}$$

If necessary, repeat steps 4 through 7 until the estimates of α_{X_i} converge with a preassigned tolerance. A tolerance level of 0.005 is common. Once the direction cosines converge, the new checking point can be estimated keeping β as the unknown parameter. This additional computation may improve the robustness of the algorithm. The assumption of an initial value of β in step 2 is necessary only for the sake of this step. Otherwise, step 2 can be omitted.

- Step 8: Compute an updated value for β using the condition that the limit state equation must satisfy the new checking point.
- Step 9: Repeat steps 3–8 until β converges to a preassigned tolerance level. A low tolerance level of 0.001 can be used particularly if the algorithm is developed in a computer environment.

The algorithm converges very rapidly, most of the time within five cycles, depending upon the nonlinearity in the limit state equation. A computer program can be easily developed to implement the concept. If the problem of interest contains correlated RVs X_i's, just after defining the limit state function in Step 1, it needs to be redefined in terms of uncorrelated variables Y_i's, as discussed in Section 2.10.

Example 2.1 Haldar and Mahadevan (2000a) presented an example to clarify and discuss all nine steps discussed in the previous section. The same example, with several modifications, is considered here. They estimated the reliability index using a strength-based limit state function. However, here a mechanically equivalent stress-based limit state function is used to demonstrate that the same reliability information can be obtained. It will establish one of many attractive features of FORM. All the numbers given in the example are expressed in the SI or metric system.

Suppose a W16 × 31 steel section made of A36 steel is suggested to carry an applied deterministic bending moment of 128.80 kN-m. The nominal yield stress F_y of the steel is 248.2×10^3 kPa and the nominal plastic modulus of the section Z is 8.849×10^{-4} m³. Assume F_y is a lognormal variable with a mean value of 262×10^3 kPa and a COV of 0.1, and Z is a normal random variable with a mean of 8.849×10^{-4} m³ and a COV of 0.05. The limit state function considered by Haldar and Mahadevan (2000a) is redefined in the SI system as $F_y Z - 128.80 = 0$. The same limit state function is expressed in stress formulation and can be expressed as $F_y - \dfrac{128.8}{Z} = 0$. All the steps required to estimate the reliability index β and the corresponding probability of failure are summarized in Table 2.1.

If interested, please practice how all nine steps are evaluated at the first iteration to implement FORM Method 1 and compare the numbers given in the table and the numbers you calculated. Since the two problems are mechanically equivalent, the estimation of the final reliability index obtained by the strength and stress formulations are identical, as expected, although intermediate steps are slightly different.

2.11.2 Correlated Non-Normal Variables

In general, reliability evaluation for limit state equations with correlated non-normal variables is quite involved. Since FORM is an iterative process, at each iteration the checking point and the corresponding equivalent mean and standard deviation of non-normal variables are expected to be different. The limit state equation needs to be redefined at each iteration by transforming correlated non-normal X_i's variables to uncorrelated normal variables Y_i's. Generally, computer programs are used to estimate the reliability index using FORM when variables are correlated. The same problem was solved, and the reliability indexes

Table 2.1 Steps in FORM Method 1.

	C1	C2	C3	C4	C5	C6	C7	C8
Step 1	$g(\ ) = F_y - (128.80)/Z$							
Step 2 β	3.0			5.01473		5.1473		
Step 3 f_y^*	2.62000E5	1.85167E5	1.97568E5	1.97783E5		1.61047E5	1.66630E5	1.67013E5
z^*	8.84901E-4	8.49293E-4	8.32699E-4	8.33811E-4		7.99773E-4	7.73281E-4	7.71197E-4
Step 4 $\mu_{F_y}^{N}$	2.606965E5	2.48515E5	2.52351E5	2.52410E5		2.38618E5	2.41212E5	2.41383E5
$\sigma_{F_y}^{N}$	2.613448E4	1.84706E4	1.97076E4	1.972917E4		1.60646E4	1.65883E4	1.66597E4
μ_z^{N}	8.84901E-4	8.84901E-4	8.84901E-4	8.84901E-4		8.84901E-4	8.84901E-4	8.84901E-4
σ_z^{N}	4.42451E-5	4.42451E-5	4.42451E-5	4.42451E-5		4.42451E-5	4.42451E-5	4.42451E-5
Step 5 $\left(\dfrac{\partial g}{\partial F_y}\right)^{*}$	1.00000	1.00000	1.00000	1.00000		1.00000	1.00000	1.00000
$\left(\dfrac{\partial g}{\partial z}\right)^{*}$	1.6448507E8	1.7856664E8	1.8575475E8	1.8525948E8	2.0136408E8	2.1318239E8	2.1539785E8	2.1656326E8
Step 6 α_{F_y}	0.96335	0.91942	0.92296	0.92347	0.874	0.87166	0.86709	0.86685
α_z	0.26826	0.39328	0.38490	0.38367	0.486	0.49012	0.49816	0.49857
Step 7	Go to Step 3. Compute the new checking point using information from Step 6.							
Step 8 β	5.0147	5.1587	5.1473	5.1473			5.1587	5.1509
Step 9	Repeat Steps 3 through 8 until β converges.							

were estimated for the correlated normal variables and the correlated non-normal variables cases in Haldar and Mahadevan (2000a). The reliability indexes when both variables are correlated normal and one of them is normal and the other is lognormal are estimated to be 3.922 and 4.586, respectively. In both cases, the correlation coefficients between F_y and Z are assumed to be 0.3. It was 5.151 when the variables were uncorrelated and one of them is normal and the other is lognormal. Consideration of correlation characteristics of RVs is very important in reliability estimation.

2.12 Probabilistic Sensitivity Indices

All RVs in a problem do not have equal influence on the estimated reliability index. A measure called the sensitivity index can be used to quantify the influence of each basic RV. A gradient vector, denoted as $\nabla g(\mathbf{Y})$, containing the gradients of the limit state function in the space of standard normal variables, can be used for this purpose. In the standard normal variable space, the mean and standard deviation of Y_i's are zero and unity, respectively. Let $\boldsymbol{\alpha}$ be a unit vector in the direction of this gradient vector. The design point can be shown to be $\mathbf{y}^* = -\beta\,\boldsymbol{\alpha}$, and

$$\alpha_i = -\frac{\partial\beta}{\partial y_i^*} \tag{2.60}$$

This equation states that the elements of the vector $\boldsymbol{\alpha}$ are directly related to the derivatives of β with respect to the standard normal variables. If these are related to the variables in the original space and using their statistical properties, a unit sensitivity vector can be derived as (Der Kiureghian and Ke 1985):

$$\gamma = \frac{\mathbf{S}\,\mathbf{B}^t\boldsymbol{\alpha}}{|\mathbf{S}\,\mathbf{B}^t\boldsymbol{\alpha}|} \tag{2.61}$$

where $\mathbf{S}$ is a diagonal matrix of standard deviations of input variables (equivalent normal standard deviations for the non-normal variables) and $\mathbf{B}$ is also a diagonal matrix required to transform the original variables $\mathbf{X}$ to equivalent uncorrelated standard normal variables $\mathbf{Y}$, i.e. $\mathbf{Y} = \mathbf{A} + \mathbf{B}\,\mathbf{X}$. For the ith RV, this transformation is $Y_i = (X_i - \mu_{X_i})/\sigma_{X_i}$. Thus, matrix $\mathbf{B}$ contains the inverse of the standard deviations or the equivalent normal standard deviations. If the variables are uncorrelated, then in Eq. 2.61, the product $\mathbf{S}\,\mathbf{B}^t$ will be a unit diagonal matrix. Thus, the sensitivity index vector will be identical to the direction cosines vector of the RVs. If the variables are correlated, the transformation matrix $\mathbf{T}$, as in Eq. 2.56, needs to be considered. Then, the sensitivity index vector and the direction cosines vector will be different.

The elements of the vector γ can be referred to as sensitivity indices of individual variables. The information on the sensitivity indices can be used to improve the computational efficiency of the algorithm. Variables with low sensitivity indices at the end of the first few iterations can be treated as deterministic at their mean values for the subsequent iterations in the search for the minimum distance. This additional step is expected to significantly reduce the computational efforts since as a practical matter, only a few variables will have a significant effect on the probability of failure. Since the reliability evaluation of nonlinear dynamic systems excited in the time domain is expected to be extremely challenging and demanding, the reduction in the size of the problems will always be attempted to increase the efficiency of an algorithm without compromising the accuracy. It will be discussed in more detail in Chapter 5 in developing the REDSET concept.

2.13 FORM Method 2

Using Step 8 discussed in Section 2.11.1, the explicit limit state equation needs to be solved to update the information on the reliability index and then the coordinates of the new checking points in FORM Method 1. This can be difficult in most cases of practical interest for complicated nonlinear limit state functions. In some cases, the explicit expression for functions may not even be available. This will limit the usefulness of FORM Method 1. An alternative Newton–Raphson recursive algorithm, referred to as FORM Method 2 in Haldar and Mahadevan (2000a) can be used. It is similar to FORM Method 1 in that it linearizes the limit state function at each checking point; however, instead of solving the limit state equation explicitly for β, it uses the derivatives to find the coordinates of the next iteration point. As discussed earlier, FORM Method 2 will not be of interest to develop REDSET, the major thrust of this book and will not be considered further.

2.14 System Reliability Evaluation

FORM-based algorithms generally estimate the probability of failure for a single-limit state equation for a specific structural element. It needs to be used several times to study multiple performance criteria and to consider failures of multiple elements. The concept used to consider multiple failure modes and/or multiple component failures is generally known as the system reliability evaluation. A complete reliability analysis includes both the component-level and system-level estimates.

Two basic approaches used for the system reliability evaluation are the failure mode approach (FMA) and the stable configuration approach (SCA). In the FMA approach, all possible ways a structure can fail are identified using a fault tree diagram. FMA is very effective for systems with ductile components. The SCA approach considers how a system or its damaged states can carry loads without failure. SCA is effective for highly redundant systems with brittle components or with ductile and brittle components.

In general, system reliability evaluation involves evaluating the probability of union and intersection of events considering the statistical correlation between them. Since statistical correlations are difficult to evaluate, the upper and lower bounds are generally estimated for the system reliability evaluation. These bounds, generally known as the first-order bounds, are usually estimated by assuming all the events are either perfectly correlated or statistically independent. If the first-order bounds are very wide, second-order bounds can be estimated. Haldar and Mahadevan (2000a) commented that these bounds are not true bounds of the system failure probability.

In any case, when a system is represented by finite elements, as will be the case for this book, the information on the probability of failure will always be generated using the overall system behavior, no system evaluation is necessary. The structural stiffness matrix will be redefined in the case of system failure caused by the failure of one or multiple structural elements in strength. The system will also develop unacceptable large deformation, more than the allowable value when it fails to satisfy the serviceability requirement.

2.15 Fundamentals of Monte Carlo Simulation Technique

In the previous sections, various methods with different levels of difficulty and sophistication are discussed to estimate the reliability index or safety index to estimate the probability of failure. Most of these methods require that the limit state equations are available in the explicit form and a reasonable background in probability and statistics is required, more than one undergraduate course, to extract the reliability information. Since all undergraduate students in civil engineering will have the necessary background, at least in the United States, they are expected to estimate the underlying reliability information. Students from other engineering disciplines or professionals from older generations may not have the necessary exposure to these areas to implement FORM or other similar procedures. However, the most recent trend indicates that professionals in all disciplines may have to estimate the underlying risk before making important decisions. They may not be familiar with the process of making a decision in the presence of uncertainty in a problem. However, most of them are expected to have limited exposure to the basic probability and statistics concepts: mean, variance, few commonly used distributions, etc. They are also very comfortable in using very advanced computational power currently available in numerous computer applications. They use simulation, knowingly or unknowingly, in many different applications and will be open to estimating the probability of failure of engineering systems with little guidance. In fact, simulation is commonly used by experts to independently evaluate the accuracy and efficiency of any new reliability evaluation technique to incorporate uncertainty in the formulation. Simulation has now become an integral part of any basic research.

The basic concept of simulation in the simplest form can be discussed as follows. All RVs in a problem can be sampled numerous times according to their PDFs indicating uncertainty in them. Considering each realization of all the RVs in the problem produces a set of numbers that indicates one realization of the problem itself. Solving the problem deterministically for each realization is known as a simulation trial, run, or cycle. Using a large number of simulation cycles is expected to give an overall probabilistic characteristic of the problem, particularly when the number of cycles N tends to infinity. The simulation technique using a computer is an inexpensive way to study the implication of the presence of various sources of uncertainty in the problem as opposed to a limited number of laboratory investigations. In fact, simulation has become the building block for most forecasting-related problems including in developing the artificial intelligence algorithm. It is used to verify any new predictive model. Results obtained from limited experimental investigations are now required to be verified using simulation. Complex theoretical models also need to consider the implications of the presence of uncertainty in the formulation.

The most commonly used technique used for simulation in engineering applications is known as the Monte Carlo simulation (MCS) technique. The name Monte Carlo is associated with a place where gamblers take the risk. The name itself has no mathematical significance. It has nothing to do with gambling. It was used as code words by von Neumann during World War II for the development of nuclear weapons at the Los Alamos National Laboratory, New Mexico. With the significant advancement of computer technology, simulation has now become an important powerful and dependable tool for reliability estimation of engineering systems for both explicit and implicit limit state functions. The basic simulation concept is relatively straightforward and can be used by engineers with basic knowledge of probability and statistics for estimating the underlying risk of complicated engineering problems.

2.15.1 Steps in Numerical Experimentations Using Simulation

The basic MCS technique can be implemented with the help of six basic steps as discussed below.

- Step 1: Define the limit state function and identify all RVs in the formulation.
- Step 2: Quantify the probabilistic characteristics of all RVs in terms of their PDFs and the corresponding parameters to define them.
- Step 3: Generate N numbers of values for all RVs in the problem according to their PDFs. Most commercially available computer programs have this feature built-in. They can also be generated as suggested in Haldar and Mahadevan (2000a).
- Step 4: Deterministically evaluate the problem N times for each set of realizations of all RVs. It is commonly termed as numerical experimentation.
- Step 5: Extract the reliability information from N data points obtained from Step 4 as if they were observed or obtained experimentally, and
- Step 6: Determine the accuracy and efficiency of the generated data obtained by simulation.

The selection of N, the total number of simulations for the problem under consideration can be difficult since it will depend on the unknown probability of failure. It will depend on the experience of the simulators.

2.15.2 Extracting Probabilistic Information from *N* Data Points

In most engineering applications, the evaluation of statistical characteristics of N data points may not be the primary interest. Simulations are conducted primarily to estimate the probability of failure or to evaluate the accuracy of a prediction. Suppose, a limit state equation $g(\ )$ is represented by Eq. 2.28. A negative value of the equation will indicate failure or that it did not satisfy the requirement for which it was designed. Suppose N_f is the number of simulation cycles when $g(\ )$ is negative and N is the total number of simulation cycles. With this information, the probability of failure, p_f can be estimated as:

$$p_f = \frac{N_f}{N} \tag{2.62}$$

This equation clearly indicates that the unknown p_f will depend on the total number of simulations. This can be a major hurdle and the extracted information on risk can be open to challenge if the total number of simulations is not large enough. The related issues are discussed in more detail by Haldar and Mahadevan (2000a).

2.15.3 Accuracy and Efficiency of Simulation

In many engineering problems, the probability of failure could be of the order of 10^{-5}. Therefore, on an average, only 1 out of 100 000 trials would show a failure. This indicates that at least 100 000 simulation cycles are required for this problem.

For an acceptable reliability estimate, Haldar and Mahadevan suggested at least 10 times this value or at least one million cycles will be needed to estimate the underlying risk. If a problem has n number of RVs, then n million random numbers are necessary to estimate the probability of failure using MCS. The task is not really that complicated since everything is done by the computer. As stated earlier, with a basic understanding of the uncertainty quantification process, anyone can simulate complicated engineering problems.

In practice, one does not need to simulate a very large number of times. Numerous variance reduction techniques (VRTs) and space reduction procedures are available to extract the information on risk using very few simulations. Some of these procedures can also be combined producing compounding beneficial effects to extract the risk information using only few dozens or hundreds of simulations. However, it will require a considerable amount of expertise not expected from everyday users for routine applications. Thus, simplicity, the basic attractiveness of simulation will be lost. An alternative to simulation is required.

Simulation of correlated variables is also expected to be straightforward. The limit state function needs to be redefined in terms of uncorrelated variables as discussed in Section 2.10. Before completing the discussions on the fundamentals of reliability estimation, the following comments will be very relevant. In the very early stage of the development of the reliability-based analysis and design concept, the required limit state functions were considered to be explicit in nature. To satisfy this requirement, the risk of failure of a structural element; member of a truss, a beam, or a column was estimated. Generally, a structure is composed of multiple structural elements. Estimating the probability of failure of the structure would require estimation of the probability of failure of multiple elements in a structure. The task was very complicated and challenging. Instead, the lower and upper bounds of the probability of failure were estimated. To estimate them, a structure was modeled as a series or weakest link system, or a parallel system. For relatively simple systems, expressions for work done by the external and internal forces were used for specific collapse mechanisms to explicitly define the limit state equation and then the information on risk was extracted. They depend on the expertise of the evaluator. Thus, the accuracy and uniformity of the estimated risk cannot be assured. If the simulation is used instead, this step may not be required removing one of the important sources of uncertainty.

2.16 Concluding Remarks

Some of the basic concepts required for the risk-based analysis and design include the set theory, mathematics of probability, modeling of uncertainty, commonly used probability distributions, and estimation of their parameters, fundamentals of reliability analysis for uncorrelated and correlated variables, system reliability, and the concept behind the simulation technique are briefly reviewed in this chapter. The information is expected to give sufficient knowledge to follow all the discussions made later in this book.

In spite of significant advancements made in the overall understanding of risk assessment and management, major knowledge gaps still exist in a few related areas. Reliability estimations when the limit state functions are implicit and engineering systems are excited dynamically in the time domain are some of the areas that need additional attention from the engineering profession. To understand some of the advanced concepts discussed in the book, the readers are encouraged to refer to the book by Haldar and Mahadevan (2000a), also published by Wiley. In this book, numerous examples and exercise problems are given for a broader understanding of the subject. The book is used very widely in the world. The book should be used to clarify any questions that may arise.

3

Implicit Performance or Limit State Functions

3.1 Introductory Comments

Risk or reliability is always estimated with respect to a performance or limit state function (LSF). Extraction of reliability information using FORM becomes relatively simple if an LSF of interest is available in an explicit form in terms of resistance and load-related random variables (RVs) and the performance requirement, i.e. allowable or permissible value. In the context of this book, for nonlinear dynamic problems excited in the time domain and structures represented by finite elements, the required LSFs are not expected to be implicit in nature; they may be available in an algorithmic form. For a wider discussion on LSF, it will be beneficial to break it down into two components: (i) allowable or permissible response, and (ii) structural response caused by the excitation. It can be mathematically represented as:

$$Z = \delta_{\text{allow}} - g(X_1, X_2, ..., X_n) \tag{3.1}$$

where δ_{allow} is the allowable or permissible value generally suggested in design guidelines and codes or specified by owners/users/designers to satisfy the strength and serviceability requirements for the intended use of the structure, and $X_1, X_2, ...,$ X_n are load- and resistance-related RVs present in the formulation to estimate the structural response caused by the excitation. The failure event can be described as $Z < 0$ or when the structural response is greater than δ_{allow}. For *complex nonlinear dynamic engineering systems* (CNDES) excited in the time domain, the responses are expected to be a function of time and cannot be expressed explicitly in terms of all the RVs present in the formulation. For the reliability estimation, the maximum response information will be of interest. For the general discussion, $g(X_1,$ $X_2, ..., X_n)$ will be denoted hereafter as the response surface (RS), $g(\mathbf{X})$. The LSF of interest will be defined as $Z = 0$ by incorporating the information on δ_{allow}.

Reliability Evaluation of Dynamic Systems Excited in Time Domain: Alternative to Random Vibration and Simulation, First Edition. Achintya Haldar, Hamoon Azizsoltani, J. Ramon Gaxiola-Camacho, Sayyed Mohsen Vazirizade, and Jungwon Huh.
© 2023 John Wiley & Sons, Inc. Published 2023 by John Wiley & Sons, Inc.

It defines the boundary between the less than and more than the permissible value and represents a state beyond which a structural element or system can no longer satisfy the need for which it was built. If the load is applied statically, even for the implicit RS, a Newton–Raphson type recursive algorithm, refer to as FORM Method 2 as briefly mentioned in Chapter 2, can be used. However, if the load is applied dynamically, the RS will be implicit, and no method is currently available to extract the reliability information. Reliability estimation for the implicit LSF has been a major challenge to the profession for a long period of time. This is comprehensively discussed in this chapter. Several novel risk estimation strategies for problems with implicit RSs are investigated to identify the most attractive and robust alternative.

3.2 Implicit Limit State Functions – Alternatives

Most of the LSFs for realistic structural systems are expected to be implicit since the mathematical expression for the unknown RS, $g(\mathbf{X})$ cannot be expressed in an explicit form. Since the partial derivatives of the RS with respect to the RVs will not be readily available for implicit LSFs, it will be difficult and challenging to extract reliability information using a commonly used method like FORM. Several computational approaches can be pursued for the reliability evaluation of systems with implicit LSFs. They can be broadly divided into three categories, based on their essential philosophy. They are (i) sensitivity-based analysis, (ii) Monte Carlo Simulation (MCS) including numerous efficient sampling schemes and space reduction techniques, and (iii) the RS concept. The sensitivity-based approach will not be of interest for this book. It was used in developing the SFEM concept discussed in Chapter 1. It is discussed very briefly here for the sake of completeness. Further details on it can be found in Haldar and Mahadevan (2000b).

In the implementation of FORM, the search for the design point or coordinates of the most probable failure point (MPFP) requires only the value and gradient of the RS function at each iteration. The value of the function can be obtained from the deterministic structural analysis. The gradient can be obtained using sensitivity analysis. For explicit RS functions, the gradient can be computed analytically or by conducting numerical differentiation of the function with respect to each RV in the formulation. For implicit RS functions, several approximate techniques including the finite difference, classical perturbation, and iterative perturbation methods can be used. The finite difference approach is used for the approximate numerical computation of derivatives of any function with respect to the design variables. If the analytical differentiation of the RS function $g(\mathbf{X})$ is not possible, the simplest approximate numerical approach to compute the derivative is to change an RV X by a small amount preferably close to zero and estimate the corresponding change in the value of the function. This is the basis of the finite difference approach.

In the classical perturbation, the chain rule of differentiation can be is used to compute the derivatives of structural response or RS function with respect to the basic RVs. Iterative perturbation method is suitable in the context of nonlinear structural analysis where the structural responses are obtained through an iterative process as used in implementing the SFEM concept.

Reliability analysis of a structure represented by finite elements (FEs) and excited dynamically in the time domain, for low probability events using the basic MCS technique, may require continuous running of a computer for several years. To overcome this deficiency many sophisticated space reduction or parallel processing techniques were proposed. Of course, the implementation of these techniques will require a considerable amount of expertise. Consider a typical acceleration time history of an earthquake. Suppose the total duration of an acceleration time history is 30 seconds and it is available with a time increment of 0.005 seconds. It will require total time points of (30/0.005) 6000 to show the variations of the load history with time. For a deterministic nonlinear dynamic analysis of a realistic structural system represented by FEs using a computer may require about 10 minutes to 1 hour. Assuming a deterministic analysis will require 10 minutes, for one million basic MCS will require over 19 years $[(10 \times 10^6)/(60 \times 24 \times 365) = 19.03]$ of continuous running of a computer. This simple example clearly demonstrates that the basic MCS is not expected to be a realistic risk estimation approach for CDNES considered in this book. In spite of significant advancement of computational power, an alternative to basic simulation is required for routine applications and will be proposed. Eliminating the sensitivity-based and MCS approaches from further consideration, the remaining alternative is the RS method (RSM)-based approach. Its applications for estimating structural risk may not be simple and will require a considerable amount of additional work to satisfy the reliability community. Since, the RSM-based approach is an important building block of the proposed REDSET concept, it needs to be presented more formally.

3.3 Response Surface Method

The primary purpose of response surface method (RSM) in the structural reliability analysis is to approximately construct a polynomial in explicit form to represent the implicit response function $g(\mathbf{X})$ in Eq. 3.1, where $\mathbf{X}$ is a vector containing all the load and resistance-related RVs in the response function. If the polynomial representing the structural responses is generated appropriately to satisfy the needs, the required LSF will be available in the explicit form and the basic FORM-based approach can be used to extract the reliability information in a relatively straightforward way. However, the optimum number of structural response data are needed for fitting a polynomial. The required response data can be generated at selected locations by designing a set of experiments. This will require multiple nonlinear deterministic FE-based dynamic analyses of the structure. However, if developed properly, it will be several orders of magnitude smaller than that of required for the basic MCS method. It will require integration of several advanced mathematical concepts producing compounding beneficial effects. The process will lead to the development of the REDSET concept as will be discussed in this chapter and in Chapter 5.

The basic RSM concept is essentially a set of statistical techniques assembled to find the best value of the responses considering the uncertainty or variations in the design variables (Khuri and Cornell 1996). However, the basic RSM concept was developed to address problems in biology and agriculture (Box et al. 1978). It cannot be directly used to approximately generate an implicit RS function explicitly representing the structural response data or their statistics in terms of the mean and the coefficient of variation (COV). For the structural reliability evaluation, several major improvements of the basic RSM concept are necessary to improve the quality and quantity of the response data generated using FE analyses as discussed next.

3.4 Limitations of Using the Original RSM Concept for the Structural Reliability Estimation

The basic RSM-based concept was developed for different applications and cannot be directly used for the structural reliability estimation for several reasons. The concept was developed in the coded variable space using information only on mean and standard deviation, completely ignoring the distributional information of all the RVs. This approach will be unacceptable to the structural reliability community who uses FORM for the reliability information explicitly using complete statistical information of all RVs. Furthermore, if the RS is not generated in the failure region, any information generated using it to estimate the underlying risk may not be acceptable or appropriate. It may not be possible to postulate the failure region or a set of values for all RVs that will cause failure at the initiation of the risk evaluation study for CNDES. In addition, the total number and locations of sampling points where responses need to be estimated for developing a polynomial expression for the RS of interest are some of the important but open questions that need to be addressed. For computational efficiency, a RS needs to be developed using the absolute minimum number of response data, since each sample point will require a nonlinear time domain finite element dynamic analysis of a structure. Any extra analyses will require additional computational time. Moreover, for the nonlinear dynamic analysis, an iterative strategy will be necessary. A new RS needs to be generated for each iteration adding further to the computational time.

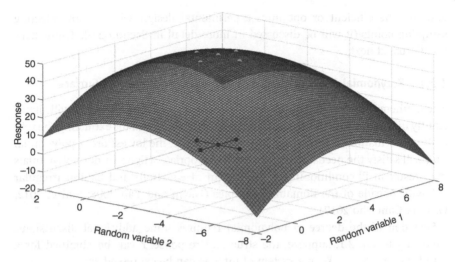

Figure 3.1 The basic concept of an IRS.

Overcoming some of these issues can be challenging. However, if all the issues identified here are appropriately mitigated, an explicit expression for the RS and the corresponding LSF will be available. Then, it will be relatively simple to extract reliability information using the commonly used and relatively simple FORM.

In summary, these discussions clearly indicate that the basic RSM concept will require significant improvements before it can be used for the structural reliability estimation. To differentiate between the original and an improved RS by removing all its deficiencies for the structural engineering applications, it will be denoted hereafter as the *improved response surface* (IRS) method. The basic concept is shown in Figure 3.1 and the processes required to generate it are discussed next.

3.5 Generation of Improved Response Surfaces

The major task at this stage is to generate appropriate and efficient IRSs in the failure region that will satisfy the reliability community using as few deterministic nonlinear FE-based dynamic time domain analyses of a structure as practicable. This task can be completed by generating deterministic response data obtained from FE analyses of the structure at sampling points around a center point following specific sampling design schemes using any available and suitable computer program and then fitting a polynomial through these points. The information on the required IRS can be generated using the following three steps. They are the selection of: (i) the degree of polynomial required to represent an IRS, (ii) the location of the center points around which sampling points will be selected, and (iii) the efficient or optimum experimental design schemes for selecting sampling points (it will be discussed in more detail in Section 5.6). These steps are discussed next.

3.5.1 Polynomial Representation of an Improved Response Surface

The mathematical form of an IRS and the processes used to generate it will significantly influence the accuracy and efficiency of an algorithm for extracting the reliability information. Considering only the efficiency, the lowest order polynomial that will satisfy the needs will be the most appropriate. Higher-order polynomials may result in ill-conditioning of the system of equations and exhibit irregular behavior outside of the domain of samples (Gavin and Yau 2008; Sudret 2012; Gaxiola-Camacho 2017).

Selection of the degree of polynomial requires some additional discussions. Referring to Eq. 3.1, suppose, the structural response Y can be obtained for a set of RVs $X_1, X_2, ..., X_k$ in a system of interest can be expressed as:

$$Y = g(\mathbf{X}) = g(X_1, X_2, ..., X_k) \tag{3.2}$$

The RS $g(\mathbf{X})$ is assumed to be a continuous function of $X_1, X_2, ..., X_k$. If the RS is continuous and smooth, it can be represented with a polynomial with any degrees of approximation with a Taylor series expansion about an arbitrary point $\mathbf{z} = (z_1, z_2, ..., z_k)$, i.e.

$$\begin{aligned} g(\mathbf{X}) = g(\mathbf{z}) &+ \sum_{i=1}^{k} \frac{\partial g}{\partial X_i}\bigg|_{\mathbf{z}} (X_i - z_i) + \frac{1}{2} \sum_{i=1}^{k} \frac{\partial^2 g}{\partial X_i^2}\bigg|_{\mathbf{z}} (X_i - z_i)^2 \\ &+ \sum_{i=1}^{k-1} \sum_{j>i}^{k} \frac{\partial^2 g}{\partial X_i X_j}\bigg|_{\mathbf{z}} (X_i - z_i)(X_j - z_j) + \mathbf{HOT} \end{aligned} \tag{3.3}$$

where HOT stands for Higher Order Terms, and the first and second derivatives, $\frac{\partial g}{\partial X_i}, \frac{\partial^2 g}{\partial X_i^2}$, and $\frac{\partial^2 g}{\partial X_i X_j}$, are evaluated at $\mathbf{z} = (z_1, z_2, ..., z_k)$. This equation can be expressed in a polynomial form as:

$$Y = g(\mathbf{X}) = b_0 + \sum_{i=1}^{k} b_i X_i + \sum_{i=1}^{k} b_{ii} X_i^2 + \sum_{i=1}^{k-1} \sum_{j>1}^{k} b_{ij} X_i X_j + \text{HOT} \tag{3.4}$$

where $X_i (i = 1, 2, ..., k)$ is the ith RV, b_0, b_i, b_{ii}, and b_{ij} are the unknown regression coefficients to be estimated to develop the polynomial, and k is the total number of RVs in the formulation. Considering only the linear terms, the first-order representation of Eq. 3.4 can be obtained as:

$$Y = g(\mathbf{X}) = b_0 + \sum_{i=1}^{k} b_i X_i + \varepsilon_1 \tag{3.5}$$

where ε_1 is a random error for neglecting the higher-order terms. The second-order representation can be expressed as:

$$Y = g(\mathbf{X}) = b_0 + \sum_{i=1}^{k} b_i X_i + \sum_{i=1}^{k} b_{ii} X_i^2 + \sum_{i=1}^{k-1} \sum_{j>1}^{k} b_{ij} X_i X_j + \varepsilon_2 \tag{3.6}$$

The second-order representation can have two forms: without and with cross terms $X_i X_j$. Equation 3.6 represents a second-order polynomial with cross terms. The second-order polynomial without cross terms can be expressed as:

$$Y = g(\mathbf{X}) = b_0 + \sum_{i=1}^{k} b_i X_i + \sum_{i=1}^{k} b_{ii} X_i^2 + \varepsilon_3 \tag{3.7}$$

The major implication of considering a second-order polynomial without and with cross terms is the total number of unknown coefficients required to be estimated using the response data generated by conducting FE analyses. It can be shown that the total number of coefficients required to represent Eqs. 3.6 and 3.7 are $(k+1)(k+2)/2$ and $(2k+1)$, respectively, indicating that the value of k will have significant impact on the computational efficiency of any algorithm as will be demonstrated with the help of examples later.

The degrees of a polynomial to be used to represent an IRS depend on the type of nonlinearity expected in the response. Since the nonlinear dynamic response information is under consideration, it is not expected to be linear and can be eliminated from further consideration. It is also known that the use of a polynomial greater than a second-order is generally not advisable since a large number of additional coefficients will be required to represent it (Khuri and Cornell 1996). In addition, if the value of k is relatively large, as expected for large realistic problems, even more number of coefficients needed to be estimated. Considering the information available in the literature on the characteristics of the polynomial, and to make the algorithm efficient and accurate, a second-order polynomial without and with cross terms is considered for CNDES under consideration here. To approximately represent actual IRSs, they will be expressed hereafter as:

$$\hat{g}(\mathbf{X}) = b_0 + \sum_{i=1}^{k} b_i X_i + \sum_{i=1}^{k} b_{ii} X_i^2 \tag{3.8}$$

and

$$\hat{g}(\mathbf{X}) = b_0 + \sum_{i=1}^{k} b_i X_i + \sum_{i=1}^{k} b_{ii} X_i^2 + \sum_{i=1}^{k-1} \sum_{j>1}^{k} b_{ij} X_i X_j \tag{3.9}$$

where $\hat{g}(\mathbf{X})$ is the approximate representation of the original unknown IRS $g(\mathbf{X})$. Estimations of all the unknown regression coefficients b_0, b_i, b_{ii}, and b_{ij} of these equations are the major tasks at this time.

3.6 Experimental Region, Coded Variables, and Center Point

Equations 3.8 and 3.9 need to be generated in the failure region. The unknown regression coefficients need to be estimated using the response data calculated by conducting a set of deterministic nonlinear dynamic FE analyses.

3.6.1 Experimental Region and Coded Variables

To estimate the coefficients in Eqs. 3.8 and 3.9, the response data need to be generated by conducting multiple deterministic time domain nonlinear FE analyses in the failure region at sampling points selected in a systemic way following a specific sampling design scheme, commonly known as the experimental sampling design scheme, around the center point. The experimental region is defined as the area or space inside the boundary made of all the outer experimental sampling points in the physical variable space. A narrower experimental region is always preferable since it gives more accurate information on the polynomial to be estimated (Khuri and Cornell 1996).

In the original RSM scheme, the information on the uncertainty in the RVs was defined in the physical variable space of the experimental sampling region without considering their distribution information. The coordinate of the ith RV for the jth experimental sampling point was defined as:

$$X_{ij} = X_i^C \pm h_i x_{ij} \sigma_{X_i} \text{ where } i = 1, 2, ..., k \quad \text{and} \quad j = 1, 2, ..., N \qquad (3.10)$$

where k is the number of RVs, N is the number of experimental sampling points, X_i^C and σ_{X_i} are the values of the center point and the standard deviation of an RV X_i, respectively, h_i is an arbitrary factor that defines the experimental region and is generally chosen in the range between 1 and 3 in the literature (Bucher and Bourgund 1990; Rajashekhar and Ellingwood 1993; Kim and Na 1997), and x_{ij} is the coded variable. It is important to note that Eq. 3.10 does not contain any information on the distribution of an RV. By rewriting Eq. 3.10, the coded variable x_{ij} can be defined as:

$$x_{ij} = \frac{X_{ij} - X_i^c}{h_i \sigma_i} \text{ where } i = 1, 2, ..., k \text{ and } j = 1, 2, ..., N \qquad (3.11)$$

From Eq. 3.11, it can be observed that the coded values at the center point will be zero for all the RVs. The equation transforms the input variables into a unitless system in the coded variable space. Since the coding eliminates the units of measurements of the input variables and denoting $\mathbf{X}$ as the matrix of sampling points in the coded variable space, it is relatively simple to form $\mathbf{X}$ or $\mathbf{X}^T\mathbf{X}$ matrix required for the estimation of the coefficients of a polynomial of interest. In general, the simpler form of $\mathbf{X}$ or $\mathbf{X}^T\mathbf{X}$ matrix makes it easier to invert increasing computational accuracy in the estimation of the coefficients and reduction in the computing time (Snee 1973). It also enhances the interpretability of the estimated coefficients in the model. The discussions indicate that appropriately selecting the experimental region and coded variables, the accuracy in the estimation of coefficients in Eqs. 3.8 and 3.9 can be considerably improved. At the same time, it also reduces the computational time increasing efficiency. At this stage, the appropriate experimental design needs to be set up to select the experimental sampling points where the response data will be generated by conducting the deterministic FE analyses. It needs to be repeated that Eq. 3.11 also completely ignores the distributional information of all RVs in the formulation.

3.6.2 Experimental Design

A design which dictates locations where response data need to be collected to estimate all the parameters in the polynomial represented by Eqs. 3.8 and 3.9 is called experimental design (Khuri and Cornell 1996). For the dynamic problem under consideration, it will indicate a set of specific points where deterministic nonlinear time domain FE analyses need to be conducted, commonly known as sampling points. The number of distinct experimental sampling points must be at least the number of coefficients in the respective polynomials.

Experimental designs in the context of the second-order polynomial are of interest for CNDES under consideration. Generally, the available designs can be divided into two categories: saturated design (SD) and classical design. A design is called saturated when the number of sampling points is exactly equal to the coefficients necessary to define a polynomial. In the classical design approach, responses are first calculated at experimental sampling points and then a regression analysis is carried out to formulate the RS. In this approach, one of the conceptually simpler designs is a factorial (factored factorial) design, in which the responses are estimated for each variable sampled at equal intervals. In order to fit a second-order surface for k input variables, the experimental sampling points must have at least three levels for each variable, leading to a 3^k factorial design. Box and Wilson (1951) introduced a more efficient approach, known as the central composite design (CCD), for fitting second-order surfaces. There are several other designs in the classical design category, namely, the Box–Behnken designs, Equiradial designs, Uniform shell designs, etc.

Considering several alternatives, two promising design techniques, namely SD and CCD are considered further. Their salient features are discussed in the following sections.

3.6.3 Saturated Design

Box and Draper (1971) suggested several saturated second-order designs. SD consists of only as many experimental sampling points as the total number of coefficients necessary to define a polynomial (Bucher et al. 1989; Bucher and Bourgund 1990; Rajashekhar and Ellingwood 1993). The unknown coefficients are obtained by solving a set of linear equations, without using any regression analysis. The design can be used for polynomials without and with cross terms. For the polynomial without cross terms, the function $g(\mathbf{X})$ represents the original function along the coordinate axes and at a center point and the design consists of one center point and $2k$ star points (one at $+1$ and one at -1 in the coded space for each variable). The total number of experimental sampling points required for this design can be shown to be $2k + 1$, where k is the total number of RVs. For the polynomial with cross terms, the design needs additional experimental sampling points of $k(k-1)/2$ for the cross terms. The total number of experimental sampling points, N, required for this design can be shown to be $N = (k+1)(k+2)/2$. In this design, the surface fits exactly at the experimental sampling points and SD needs much fewer experimental sampling points than CCD, as will be discussed in more detail later, making this the most efficient design. However, this design lacks the statistical properties (rotatability, orthogonality, etc.) and the sample space between the axes of RVs may not be covered with sufficient accuracy, causing inaccuracy in the generated IRS.

3.6.4 Central Composite Design

CCD consists of a complete 2^k factorial design, n_0 center points ($n_0 \geq 1$), and two axial points on the axis of each RV at a distance h_i in Eq. 3.10 from the design center point where $h_i = \sqrt[4]{2^k}$ in order to make the design rotatable. This design can only be used for the polynomials with cross terms. The total number of experimental sampling points, N, required for this design can be shown to be $N = 2^k + 2k + n_0$. By comparing the total number of experimental sample points required to implement SD and CCD, it is obvious that CCD will require much more than SD. CCD design is expected to be more accurate than SD and contains several desirable statistical properties, including the ability to estimate the experimental error and to test the adequacy of the model by conducting the analysis of variance (ANOVA), orthogonality, and rotatability. For ready reference, ANOVA is briefly reviewed next.

3.7 Analysis of Variance

To implement CCD, regression analysis is required to generate the required IRS. The adequacy of a selected regression model in representing the relationship between the regressor and the response variables can be studied in many ways including the estimation of the coefficient of determination R^2 with the help of ANOVA. The total variation in N experimental sampling data is denoted as the total sum of squares (SST). It has two components: the amount of variability accounted for by the regression model (SSR) and the residual variation left unexplained by the regression model (SSE). Mathematically, the concept can be expressed as:

$$SST = SSR + SSE \tag{3.12a}$$

or,

$$\sum_{i=1}^{N}(y_i - \bar{y})^2 = \sum_{i=1}^{N}(\hat{y}_i - \bar{y})^2 + \sum_{i=1}^{N}(y_i - \hat{y}_i)^2 \tag{3.12b}$$

SST can be computed as:

$$SST = \mathbf{Y}^T\mathbf{Y} - \frac{(\mathbf{1}^T\mathbf{Y})^2}{N} \tag{3.13}$$

where $\mathbf{Y}$ is a vector of size $N \times 1$ containing information on the responses and $\mathbf{1}^T$ is a unit vector.

SSR can be estimated as:

$$SSR = \hat{\mathbf{Y}}^T\mathbf{Y} - \frac{(\mathbf{1}^T\mathbf{Y})^2}{N} \tag{3.14}$$

where $\hat{\mathbf{Y}}$ is a vector of size $N \times 1$ containing information on the response values from the fitted surface. If the fitted response model contains p coefficients, then the number of degrees of freedom associated with SSR is $(p - 1)$.

The SSE can be computed as:

$$SSE = \mathbf{Y}^T\mathbf{Y} - \hat{\mathbf{Y}}^T\mathbf{Y} \tag{3.15}$$

The number of degrees of freedom for *SSE* is defined as $(N - p)$ which is the difference $(N - 1) - (p - 1) = (N - p)$. The coefficient of determination, R^2, a non-dimensional term indicating the amount of variation addressed by the regression model, can be defined as:

$$R^2 = \frac{SSR}{SST} \tag{3.16}$$

It is a measure of the proportion of total variation of $\mathbf{Y}$ about its mean explained by the selected response model. R^2 will have a value between 0 and 1. When it is close to 1, it implies that most of the variability in $\mathbf{Y}$ is explained by the regression model. The R^2 value also depends on the range of the regressor variable. It will increase as the spread of the regressor variable increases. In any case, the R^2 value gives an indication of the adequacy of the regression model used.

3.8 Experimental Design for Second-Order Polynomial

Before selecting a specific sampling scheme according to an experimental design for fitting a second-order polynomial, the characteristics of SD and CCD need further discussions. Based on the previous discussions on SD and CCD, three experimental designs are: (i) SD with second-order polynomial without cross terms, (ii) SD with a second-order polynomial with cross terms, and (iii) CCD with a second-order polynomial with cross terms. CCD without cross terms is not possible. They will be denoted hereafter as basic factorial designs in the following discussions.

3.8.1 Experimental Design – Model 1: SD with Second-Order Polynomial without Cross Terms

For three RVs, i.e. $k = 3$, the basic sampling scheme for experimental design Model 1 is shown in Figure 3.2. The total number of sampling points required to implement this model is $2k + 1$.

For Model 1, the original performance function represented as $g(\mathbf{X})$ in Eq. 3.8 can be expressed as:

$$\mathbf{Y} = \mathbf{Xb} + \varepsilon \tag{3.17}$$

where $\mathbf{Y}$ is the vector containing actual response information, $\mathbf{X}$ represents the matrix of sampling points in coded values, as shown in Table 3.1, $\mathbf{b}$ is the vector of unknown coefficients, and ε is the vector of random errors. The unknown coefficients denoted as b_0, b_i, and b_{ii} in Eq. 3.8 can be obtained by solving a set of simultaneous equations represented as:

$$\mathbf{b} = \mathbf{X}^{-1}\mathbf{Y} \tag{3.18}$$

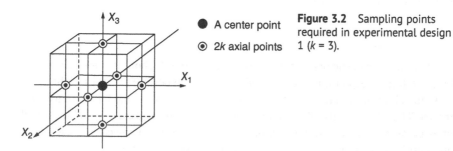

● A center point

◉ $2k$ axial points

Figure 3.2 Sampling points required in experimental design 1 ($k = 3$).

Table 3.1 Representation of **X** using experimental design – Model 1.

	x_0	x_1	x_2	...	x_{k-1}	x_k	x_1^2	x_2^2	...	x_{k-1}^2	x_k
1	+1	−1	0	...	0	0	+1	0	...	0	0
2	+1	+1	0	...	0	0	+1	0	...	0	0
3	+1	0	−1	...	0	0	0	+1	...	0	0
4	+1	0	+1	...	0	0	0	+1	...	0	0
⋮	⋮	⋮	⋮	...	⋮	⋮	⋮	⋮	...	⋮	⋮
$2k-3$	+1	0	0	...	−1	0	0	0	...	+1	0
$2k-2$	+1	0	0	...	+1	0	0	0	...	+1	0
$2k-1$	+1	0	0	...	0	−1	0	0	...	0	+1
$2k$	+1	0	0	...	0	+1	0	0	...	0	+1
$2k+1$	+1	0	0	...	0	0	0	0	...	0	0

3.8.2 Experimental Design – Model 2: SD with Second-Order Polynomial with Cross Terms

The accuracy of Model 1 can be improved by considering the cross terms in generating the RSs. The total experimental sampling points required are exactly the same as the unknown coefficients in the polynomial. This design consists of one center point, $2k$ axial points (one at +1 and one at −1 for each variable in coded values), and $k(k-1)/2$ edge points. An edge point can be defined as a k-vector having ones in the ith and jth locations and zeros elsewhere. Hence, the total number of sampling points required in Experimental Design Model 2 are $2k+1+k(k-1)/2$ or $(k+1)(k+2)/2$. The basic sampling scheme for this design model is shown in Figure 3.3 for $k = 3$.

For Model 2, the original performance function represented as $g(\mathbf{X})$, can be expressed by Eq. 3.9. The experimental sampling points in coded values are summarized as matrix **X** in Table 3.2. The unknown coefficients denoted as b_0, b_i, b_{ii}, and b_{ij} in Eq. 3.9 can be obtained by Eq. 3.18.

Figure 3.3 Sampling points required in experimental design – Model 2 ($k = 3$).

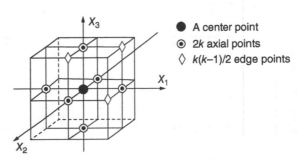

● A center point
⊙ $2k$ axial points
◊ $k(k-1)/2$ edge points

Table 3.2 Representation of **X** matrix in coded values – Model 2.

	x_0	x_1	x_2	$\cdots$	x_{k-1}	x_k	x_1^2	x_2^2	$\cdots$	x_{k-1}^2	x_k^2	$x_{1,2}$	$x_{1,3}$	$\cdots$	$x_{1,k}$	$x_{2,3}$	$\cdots$	$x_{k-1,k}$
1	+1	−1	0	$\cdots$	0	0	+1	0	$\cdots$	0	0	0	0	$\cdots$	0	0	$\cdots$	0
2	+1	+1	0	$\cdots$	0	0	+1	0	$\cdots$	0	0	0	0	$\cdots$	0	0	$\cdots$	0
3	+1	0	−1	$\cdots$	0	0	0	+1	$\cdots$	0	0	0	0	$\cdots$	0	0	$\cdots$	0
4	+1	0	+1	$\cdots$	0	0	0	+1	$\cdots$	0	0	0	0	$\cdots$	0	0	$\cdots$	0
..	$\vdots$	$\vdots$	$\vdots$	$\vdots$	$\vdots$	$\vdots$	$\vdots$	$\vdots$	$\vdots$	$\vdots$	$\vdots$	$\vdots$	$\vdots$	$\vdots$	$\vdots$	$\vdots$	$\vdots$	$\vdots$
$2k-3$	+1	0	0	$\cdots$	−1	0	0	0	$\cdots$	+1	0	0	0	$\cdots$	0	0	$\cdots$	0
$2k-2$	+1	0	0	$\cdots$	+1	0	0	0	$\cdots$	+1	0	0	0	$\cdots$	0	0	$\cdots$	0
$2k-1$	+1	0	0	$\cdots$	0	−1	0	0	$\cdots$	0	+1	0	0	$\cdots$	0	0	$\cdots$	0
$2k$	+1	0	0	$\cdots$	0	+1	0	0	$\cdots$	0	+1	0	0	$\cdots$	0	0	$\cdots$	0
$2k+1$	+1	+1	+1	$\cdots$	0	0	+1	+1	$\cdots$	0	0	+1	0	$\cdots$	0	0	$\cdots$	0
$2k+2$	+1	+1	0	$\cdots$	0	0	+1	0	$\cdots$	0	0	0	+1	$\cdots$	0	0	$\cdots$	0
..	$\vdots$	$\vdots$	$\vdots$	$\vdots$	$\vdots$	$\vdots$	$\vdots$	$\vdots$	$\vdots$	$\vdots$	$\vdots$	$\vdots$	$\vdots$	$\vdots$	$\vdots$	$\vdots$	$\vdots$	$\vdots$
..	+1	0	+1	$\cdots$	0	0	0	+1	$\cdots$	0	0	0	0	$\cdots$	0	+1	$\cdots$	0
..	+1	+1	0	$\cdots$	0	+1	+1	0	$\cdots$	0	+1	0	0	$\cdots$	+1	0	$\cdots$	+1
..	+1	0	0	$\cdots$	+1	+1	0	0	$\cdots$	+1	+1	0	0	$\cdots$	0	0	$\cdots$	+1
$(k+1)(k+2)/2$	+1	0	0	$\cdots$	0	0	0	0	$\cdots$	0	0	0	0	$\cdots$	0	0	$\cdots$	0

3.8.3 Experimental Design – Model 3: CCD with Second-Order Polynomial with Cross Terms

As discussed earlier, CCD can only be used for a polynomial with cross terms as in Eq. 3.9. It consists of three items: (i) n_0 center points, (ii) the axial portion of the design, and (iii) the factorial portion of the design. They are shown in Figure 3.4. For n_0 center points, generally $n_0 \geq 1$ and $n_0 = 1$ if the numerical experiments are conducted. For the axial portion of the design, two axial points need to be considered on the axis of each design variable at a distance of h from the center point. The factorial portion of the design contains a complete 2^k factorial design, where the factor levels are coded to the usual -1 and $+1$ values. Combining all three parts, it can be shown that the total number of experimental sampling points N, required to implement CCD is $N = 2^k + 2k + 1$. The basic sampling scheme for experimental design Model 3 is shown in Figure 3.4 for $k = 3$. As before, the experimental sampling points of this model in coded values are represented as matrix X and are shown in Table 3.3 for k RVs in which X is a matrix of size $(N \times p)$ and N is the number of experimental sampling points and p is the number of unknown coefficients.

For Model 3, the least squares estimate of $\hat{b}$, denoted by b_0, b_i, b_{ii}, and b_{ij} in Eq. 3.9, can be obtained by conducting the least squares-based regression analysis. Haldar and Mahadevan (2000a) showed that $\hat{b}$ can be estimated as:

$$\hat{b} = [X^t X]^{-1} [X^t Y] \tag{3.19}$$

where X represents the matrix of sampling points in coded values, as shown in Table 3.3 and Y is the vector containing actual response information.

3.9 Comparisons of the Three Basic Factorial Designs

It is important at this stage to compare the three basic factorial designs discussed earlier. The total number of unknown coefficient (p) and the total number of

Figure 3.4 Sampling points required in experimental design 3 ($k = 3$).

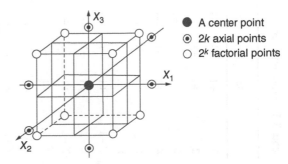

Table 3.3 Representation of X using experimental design – Model 3.

	x_0	x_1	x_2	⋯	x_{k-1}	x_k	x_1^2	x_2^2	⋯	x_{k-1}^2	x_k^2	$x_{1,2}$	$x_{1,3}$	⋯	$x_{1,k}$	$x_{2,3}$	⋯	$x_{k-1,k}$
1	+1	−1	−1	⋯	−1	−1	+1	+1	⋯	+1	+1	+1	−1	⋯	+1	+1	⋯	+1
2	+1	+1	−1	⋯	−1	−1	+1	+1	⋯	+1	+1	−1	+1	⋯	−1	+1	⋯	+1
3	+1	−1	+1	⋯	−1	−1	+1	+1	⋯	+1	+1	−1	−1	⋯	+1	−1	⋯	+1
4	+1	+1	+1	⋯	−1	−1	+1	+1	⋯	+1	+1	+1	−1	⋯	−1	−1	⋯	+1
..	⋯	⋯	⋯	⋯	⋯	⋯	⋯	⋯	⋯	⋯	⋯	⋯	⋯	⋯	⋯	⋯	⋯	⋯
..	⋯	⋯	⋯	⋯	⋯	⋯	⋯	⋯	⋯	⋯	⋯	⋯	⋯	⋯	⋯	⋯	⋯	⋯
2^k-3	+1	−1	−1	⋯	+1	+1	+1	+1	⋯	+1	+1	+1	−1	⋯	−1	−1	⋯	+1
2^k-2	+1	+1	−1	⋯	+1	+1	+1	+1	⋯	+1	+1	−1	+1	⋯	+1	+1	⋯	+1
2^k-1	+1	−1	+1	⋯	+1	+1	+1	+1	⋯	+1	+1	−1	−1	⋯	−1	−1	⋯	+1
2^k	+1	+1	+1	⋯	+1	+1	+1	+1	⋯	+1	+1	+1	+1	⋯	+1	+1	⋯	+1
2^k+1	+1	−α	0	⋯	0	0	α²	0	⋯	0	0	0	0	⋯	0	0	⋯	0
2^k+2	+1	+α	0	⋯	0	0	α²	0	⋯	0	0	0	0	⋯	0	0	⋯	0
2^k+3	+1	0	−α	⋯	0	0	0	α²	⋯	0	0	0	0	⋯	0	0	⋯	0
2^k+4	+1	0	+α	⋯	0	0	0	α²	⋯	0	0	0	0	⋯	0	0	⋯	0
..	⋯	⋯	⋯	⋯	⋯	⋯	⋯	⋯	⋯	⋯	⋯	⋯	⋯	⋯	⋯	⋯	⋯	⋯
2^k+2k-3	+1	0	0	⋯	−α	0	0	0	⋯	α²	0	0	0	⋯	0	0	⋯	0
2^k+2k-2	+1	0	0	⋯	+α	0	0	0	⋯	α²	0	0	0	⋯	0	0	⋯	0
2^k+2k-1	+1	0	0	⋯	0	−α	0	0	⋯	0	α²	0	0	⋯	0	0	⋯	0
2^k+2k	+1	0	0	⋯	0	+α	0	0	⋯	0	α²	0	0	⋯	0	0	⋯	0
2^k+2k+1	+1	0	0	⋯	0	0	0	0	⋯	0	0	0	0	⋯	0	0	⋯	0

Table 3.4 Comparison among the three experimental designs.

Experimental design	Polynomial representation	Number of unknown coefficients (p)	Number of sampling points (N)	k	p	N
Model 1 (SD)	Eq. 3.8	$2k + 1$	$2k + 1$	10	21	21
				20	41	41
				100	201	201
Model 2 (SD)	Eq. 3.9	$\frac{(k+1)(k+2)}{2}$	$\frac{(k+1)(k+2)}{2}$	10	66	66
				20	231	231
				100	5151	5151
Model 3 (CCD)	Eq. 3.9	$\frac{(k+1)(k+2)}{2}$	$2^k + 2k + 1$	10	66	1045
				20	231	1 048 617
				100	5151	≈ 1.26765 E30

sampling points required to estimate all the coefficients (N) for different numbers of RVs (k) are shown in Table 3.4.

The total number of experimental sampling points, N is directly related to the efficiency of each design since each experimental sampling point represents a deterministic time domain nonlinear FE analysis. Obviously, considering just the efficiency, Model 1 will be the most efficient since it requires the least experimental sampling points. However, it will not adequately cover the sample space between axes and may not represent the RS accurately. It has the potential for use for intermediate iterations for an iterative algorithm. Experimental Design Model 2 is expected to be more accurate than Model 1 and more efficient than Model 3. However, it lacks a few statistical properties and may not give any estimate of the experimental error and adequacy of the model representing the IRS. In order to use this design, it is necessary to assume that the experimental error is small, and that an independent estimate of the experimental error is available. Model 2 has the potential to be used in the final iteration for the problem consisting of many RVs. Model 3 cannot be implemented for most realistic problems consisting of a large number of RVs. This is definitely more accurate than the two other designs. It contains several statistical properties, such as the ability to estimate the experimental error and to test the adequacy of the model using ANOVA as discussed in Section 3.7, orthogonality that can make the calculation of ($\mathbf{X}^T\mathbf{X}$) relatively simple, and the rotatability that is used to obtain the uniform precision of the predicted value. In spite of all these merits, Model 3 cannot be implemented for

CNDES of interest. Model 3 has the potential to be used in the final iteration when the number of total RVs in the problem is relatively small, say less than 10. For realistic CNDES, in spite of its various advantages, CCD as discussed so far cannot be implemented.

3.10 Experimental Designs for Nonlinear Dynamic Problems Excited in the Time Domain

The basic experimental designs consisting of SD and CCD discussed in the previous sections may not satisfy the needs for the reliability analysis of CDNES. SD is efficient but may not satisfy the accuracy requirement. CCD will satisfy the accuracy requirement but cannot be implanted since it will require a very large number of nonlinear time domain FE analyses requiring continuous running of a computer for several years, as discussed earlier. For the reliability analysis of CDNES, new experimental designs are necessary. They can be developed by intelligently capturing the merits of SD and CCD. For an iterative algorithm for generating a second-order IRS in the failure region, SD and CCD are combined in several ways in an attempt to find the most attractive experimental design scheme as discussed next.

3.11 Selection of the Most Appropriate Experimental Design

In the context of an iterative algorithm, to incorporate efficiency without compromising accuracy, logically the most efficient model can be used for the initial and intermediate iterations to efficiently locate the failure region. Once the failure region is identified with reasonable accuracy, the most accurate model should be used to generate the required IRS in the final iteration to extract the reliability information with confidence. Following this thought process, several promising new designs were investigated by mixing the three basic factorial designs. Three of them are listed in Table 3.5.

Table 3.5 Three most promising schemes for the experimental design.

Design schemes	Initial and intermediate iterations	Final iteration
Scheme 1	Model 1	Model 1
Scheme 2	Model 1	Model 2
Scheme 3	Model 1	Model 3

In investigating the new design schemes for the reliability estimation of CDNES, several factors were considered including efficiency, accuracy, robustness, overall size of the problems, and different types of dynamic excitation applied in the time domain. Experience gained during this phase of the study also indicated that mixing the three basic factorial designs would not be enough for the reliability estimation of CDNES using REDSET. In fact, the sampling schemes shown in the table need to be significantly modified. Scheme 3 appears to be the most attractive, however, since CCD needs to be used in the last iteration, it may not be implementable for large realistic systems. Several other schemes were proposed and investigated in the context of the implementation of REDSET. They will be discussed in more detail in Chapter 5. At this stage, it will be assumed that an acceptable design scheme is now available. Before moving forward, issues related to the selection of center point need additional discussions.

3.12 Selection of Center Point

After selecting the most attractive experimental design scheme, the next important issue will be the selection of the center point around which samples will be collected following a specific design scheme. As discussed earlier, to accurately estimate the underlying risk, an IRS needs to be generated in the failure region, at least in the final iteration and the experimental region should be kept as narrow as possible. However, it will be very challenging to identify the failure region for CDNES excited in the time domain. A trial and error approach can be used but its success will depend on the experience and expertise of the simulator (Yao and Wen 1996). This will not make the algorithm robust enough for routine applications in everyday use.

The robustness of an algorithm will depend on the intelligent selection of the center point around which experimental sample points will be selected for the FE analyses to estimate structural responses. The following observation guided the authors to decide how to select the location of the center point. FORM iteratively locates the coordinates of the most probable failure point very efficiently explicitly considering the statistical information of all RVs. This observation prompted to explore if an IRS generating scheme can be integrated with the iterative process of FORM. It appears to be a reasonably straightforward approach. As discussed in Chapter 2, in the absence of any information on the location of the most probable failure point, the first iteration of FORM is routinely initiated at the mean values of all RVs. The same concept can be used in initiating the iterative process of developing an IRS. Essentially, the mean values of all RVs will the coordinates of the first center point. After generating the first IRS, the information on the coordinates of the checking point can be updated in subsequent iterations providing information on the coordinates of the next center point. This procedure will assure that the last IRS will be generated in the failure region. The integration process will be disused in more detail in Chapter 5 in the context of developing REDSET with the help of many examples.

3.13 Generation of Limit State Functions for Routine Design

Although not fully developed yet, the feasibility of approximately generating an IRS explicitly for CDNES excited in the time domain in the failure region has been established, based on the discussions made in this chapter. The last step in defining the explicit expression for the implicit LSF will be to document the process. The LSF of interest defined by Eq. 3.1 can be generated by integrating the information on IRS with the corresponding allowable or permissible values. Commonly used LSFs for structural engineering applications can be broadly divided into two groups: (i) serviceability and (ii) strength. They need to be considered separately since a structure can develop unacceptable lateral deflection or inter-story drift, or one or several components of a structure can fail in strength causing local and/or global failure. The REDSET algorithm is capable of estimating risk for both types of LSFs indicating its robustness. They are briefly discussed next.

3.13.1 Serviceability Limit State

The design of structures excited by the seismic, wave, or any other types of lateral dynamic loadings may be controlled by the serviceability-related functional requirements, e.g. inter-story drift or the overall lateral displacement. The serviceability LSF can be mathematically expressed as:

$$g(\mathbf{X}) = \delta_{\text{allow}} - \hat{g}(\mathbf{X}) \tag{3.20}$$

where δ_{allow} is the permissible or allowable deflection for the inter-story drift or overall lateral displacement, $\hat{g}(\mathbf{X})$ is the IRS generated for inter-story drift or overall lateral displacement at the location of interest, and $\mathbf{X}$ is a vector representing all the RVs present in the formulation. The allowable value for δ_{allow} is generally suggested in the design guidelines. In the absence of any such guidelines, commonly used values practiced in the region can be used. Once the required serviceability LSF is available in an explicit form, FORM can be used to estimate the reliability index and the corresponding probability of failure.

The discussion on the serviceability limit states will be incomplete without mentioning an upcoming new design concept known as *Performance-Based Seismic Design* (PBSD). It will be discussed in detail in Chapter 8. To implement the PBSD concept, δ_{allow} values will depend on the different levels of performance being considered including immediate occupancy (IO), life safety (LS), and collapse prevention (CP). These values are suggested in PBSD guidelines, specifically for steel structures at present. The required IRS will be represented by $\hat{g}(\mathbf{X})$ as before. All related topics will be discussed in Chapter 8.

3.13.2 Strength Limit State Functions

The reliability estimation for the strength LSF is more challenging. Generally, a structural element is designed for the critical design load acting on it. For example, a beam is designed for the maximum bending moment acting on it and it will define the strength LSF for the beam. For large structural systems consisting of many structural elements, the failure of one element may produce local failure without causing the failure of the system. The concept to consider failures of multiple components is generally referred to as system reliability. In general, the evaluation of system reliability is quite challenging and depends on several factors. Some of the important factors are (i) the contribution of the component failure events to the failure of the system, (ii) the redundancy in the system, (iii) the post-failure behavior of a component and the rest of the system, (iv) the statistical correlation between failure events, and (v) the progressive failure of components. The system reliability concept was briefly reviewed in Chapter 2. In the context of reliability evaluation of structures represented by finite elements, the overall stiffness matrix is expected to degrade with the failure of elements one by one. After the failure of one or multiple elements, the stiffness matrix will degrade accordingly and at one point, the structure will be unable to carry loads causing the system failure. For the reliability analysis with respect to strength LSFs, the failure of the most critical element is generally considered as the failure of the system since other components may fail in rapid successions. This concept is used here also for the reliability evaluation of strength-related LSFs.

3.13.3 Interaction Equations for the Strength Limit State Functions

Most of the elements in structural systems are represented by beam-column elements, i.e. they are subjected to both the axial load and bending moment at the same time. Interaction equations are generally used to consider the combined effect of the axial load and bending moment. In the United States, the interaction equations suggested by the American Institute of Steel Construction's (AISC's) *Load and Resistance Factor Design* (AISC 2011) manual, for two-dimensional structures are:

a) When $\dfrac{P_r}{P_c} \geq 0.2$

$$\left(\frac{P_r}{P_c} + \frac{8}{9} \frac{M_{rx}}{M_{cx}} \right) \leq 1.0 \tag{3.21}$$

b) When $\dfrac{P_r}{P_c} < 0.2$

$$\left(\frac{P_r}{2P_c} + \frac{M_{rx}}{M_{cx}}\right) \leq 1.0 \tag{3.22}$$

where P_r is the required axial strength, P_c is the available axial strength, M_{rx} is the required flexural strength about the X or strong axis, and M_{cx} is the available flexural strength about the X or strong axis.

According to AISC (2011), P_c and M_{cx} can be calculated as:

$$P_c = \varphi_c P_n \tag{3.23}$$

where $\varphi_c = 0.9$ and P_n is the axial strength and needs to be calculated using Chapter E – Design of Members for Compression presented in this design manual. and

$$M_{cx} = \varphi_b M_{nx} \tag{3.24}$$

where $\varphi_b = 0.9$ and M_{nx} is the design flexural strength about the X or strong axis and needs to be calculated using Chapter F – Design of Members for Flexure presented in the same design manual.

For two-dimensional structures, the strength limit state can be represented as:

$$g(X) = 1.0 - \left(\frac{P_r}{P_c} + \frac{8}{9}\frac{M_{rx}}{M_{cx}}\right) = 1.0 - [\hat{g}_P(X) + \hat{g}_M(X)]; \quad \text{if } \frac{P_r}{P_c} \geq 0.2 \tag{3.25}$$

$$g(X) = 1.0 - \left(\frac{P_r}{2P_c} + \frac{M_{rx}}{M_{cx}}\right) = 1.0 - [\hat{g}_P(X) + \hat{g}_M(X)]; \quad \text{if } \frac{P_r}{P_c} < 0.2 \tag{3.26}$$

where $\hat{g}_P(X)$ is the IRS corresponding to axial load, and $\hat{g}_M(X)$ is the IRS corresponding to bending moment. All other variables were defined earlier.

3.13.4 Dynamic Effect in Interaction Equations

The interaction equations used for static problems cannot be used for dynamic loadings. In dynamic analysis, P_r and M_{rx} are function of time and their maximum value may not occur at the same time. Furthermore, the axial and bending moment capacity or strength of the members, P_c and M_{cx} are also functions of time since the load effects change from compression to tension and vice versa in dynamic loading. The ratio P_r/P_c is a functions of time and their maximum values do not occur at the same time. It is also difficult to predict which interaction equation, Eq. 3.25 or Eq. 3.26, is applicable, since the P_r/P_c ratio is unknown. To address this issue, the interaction equations are divided into two parts: the part containing the effect of axial load only and the part containing the effect of bending moment only. In order to follow the intent of the design guidelines, the effects of axial load and bending moment are represented separately as $\alpha_1(P_r/P_c)$ and $\alpha_2\,(M_{rx}/M_{cx})$, respectively. The coefficients α_1 and α_2 represent the time-variant aspects of P_r/P_c and M_{rx}/M_{cx}. These modifications do not affect the efficiency or generality of the algorithm. Only an estimation of the coefficients has to be made for both polynomials.

To address this issue, the responses for each experimental sampling point are tracked, i.e. in a time domain analysis, recording the time when the interaction reaches the maximum value. This will indicate which interaction equation is appropriate for this particular sampling point. Then the contributions of $\alpha_1(P_r/P_c)$ and $\alpha_2 (M_{rx}/M_{cx})$ are evaluated for the experimental sampling point. Obviously, the effects of axial load and bending moment will be different for different sampling points, indicating the time aspect of the dynamic problem. These values can be used to generate the required IRS for the reliability analysis using FORM.

3.14 Concluding Remarks

In general, reliability estimation with implicit LSFs can be very challenging. This is one of the major reasons for the reliability estimation of CDNES excited in the time domain has not been comprehensively addressed so far. Among several alternatives discussed in this chapter, generating an implicit LSF explicitly, even approximately, using the RSM concept and the corresponding LSF appears to be very promising. And then FORM can be used to extract the reliability information. Reliability information for both serviceability and strength LSFs needs to be generated. They are discussed in this chapter.

However, the basic RSM concept was developed for different applications and cannot be used for structural reliability estimation. It requires multilevel modifications. An RS generated incorporating these modifications is denoted as an improved IRS. Several experimental designs are proposed. Considering both efficiency and efficiency, the most attractive design schemes cannot be implemented since these will require the continuous running of computers for several years. Discussions made in this chapter provide the foundation for the development of more advanced mathematical concepts producing compounding beneficial effects for estimating reliability of CDNES with implicit LSFs, one of the major impetus for developing the REDSET concept. At least nine different sampling design schemes for generating implicit LSFs will be presented in Chapter 5 in developing REDSET.

3.1.5 Concluding Remarks

4

Uncertainty Quantification of Dynamic Loadings Applied in the Time Domain

4.1 Introductory Comments

One of the major tasks for the reliability evaluation is the uncertainty quantification of all the random variables (RVs) present in the formulation. For ease of discussions, all RVs are subdivided into two groups: (i) resistance-related and (ii) load-related. In the very early stage of the development of the risk-based design concept, uncertainty in the resistance-related variables was extensively studied by collecting worldwide information from various sources and the results were published in detail (Ellingwood et al. 1980; Haldar and Mahadevan 2000a). The information is now well accepted and widely used in most technical publications on risk-based design. A simple internet search will provide the necessary information. The quantifications of uncertainties in the resistance-related variables do not need any additional discussion here. In general, the uncertainty associated with the resistance-related variables is smaller than that of dynamic loadings (earthquakes, winds, etc.). Of course, the propagation of the uncertainties in the resistance-related variables from the element to the system level in quantifying uncertainty in the response behavior can be different and difficult to predict in the presence of nonlinearity in the problem. In any case, the statement made in the literature like the total uncertainty in the capacity indicating resistance-related variables could be 40% or higher cannot be totally overlooked. The uncertainty associated with each RV will be properly identified and quantified in all the examples given in this book. The information can be used for routine applications.

Quantification of uncertainties in the load-related RVs can be subdivided into gravity and other loads. Uncertainty quantifications for the gravity loads including dead and live loads are extensively researched by reputed agencies in the United States including the National Bureau of Standards (Ellingwood et al. 1980).

Reliability Evaluation of Dynamic Systems Excited in Time Domain: Alternative to Random Vibration and Simulation, First Edition. Achintya Haldar, Hamoon Azizsoltani,
J. Ramon Gaxiola-Camacho, Sayyed Mohsen Vazirizade, and Jungwon Huh.
© 2023 John Wiley & Sons, Inc. Published 2023 by John Wiley & Sons, Inc.

The uncertainty associated with the dead load is expected to be smaller than that of the live loads. The available information on them will be provided when necessary. No additional discussion is necessary for the gravity loadings also at this stage.

Other loads in civil engineering applications include different types of lateral loadings caused by strong earthquakes and high winds. Due to the extensive use of computers, temperature-induced loading caused by heating and cooling, generally denoted as thermomechanical loading, is increasingly becoming an important loading of interest. Information on uncertainty in the earthquake and wind loadings is still evolving. Both loadings are natural phenomena, and their underlying physics is not completely understood and changes frequently as new information becomes available. Information on them cannot be explained with the help of mathematics satisfying all concerned parties. There is no unique procedure available to quantify the amount or nature of uncertainty in them. Since the information changes very frequently, it is very difficult to keep up with the changes and sometimes it requires a considerable amount of expertise to follow the available information. However, catastrophic damages caused by them are well documented. It is essential to reduce the loss of human lives and economic activities of a region caused by them. The major thrust of this discussion is to point out that these loadings are difficult to predict and predicting the underlying risk is very challenging. To mitigate the situation, information on both the underlying and acceptable risks needs to be quantified. If the risk is excessive and beyond the acceptable level, it needs to be reduced by redesigning the systems or buying insurance to cover the excess risk, as routinely practiced in the seismically active regions in the United States. Engineers are expected to play a very important role in the overall mitigation process.

Comments made by experts like the best way to design for earthquake loadings was to install a seismograph at the site indicating the scarcity of the available information to quantify uncertainty in them. It is reported that the uncertainty in the seismic loading, expressed in terms of coefficient of variation (COV), could be 100% or more indicating that it is huge and unpredictable. The three most common procedures suggested in design guidelines for seismic design in the United States are the equivalent lateral load procedure, the modal approach, and the nonlinear time domain analysis of structures. The first approach is an equivalent static approach. The second one is essentially appropriate for the linear structure. The third approach is required for the design of critical structures and is of interest in developing REDSET. Although design parameters are adjusted in the design guidelines to reflect the nature of uncertainty, it is difficult to quantify the nature of uncertainty in seismic loading.

Modeling wind loading as the flow of viscous fluid can be very simplistic. Results obtained by conducting elaborate wind tunnel testing provide the necessary information to develop the necessary design guidelines. The presence of a considerable amount of uncertainty in the test results cannot be avoided, as discussed in Chapter 1. Catastrophic damages caused by wind loading indicate that the quantification of uncertainties in it is still evolving.

Solder balls used in electronic packaging are subjected to thermomechanical loading caused by the heating and cooling of the computer. It is a different type of dynamic loading, very different from earthquake and wave loadings. Available studies in the related areas are very limited. This will be discussed in more detail in Chapter 9.

The major objective of these discussions is that the presence of a considerable amount of uncertainty in the design variable in *complex nonlinear dynamic engineering systems* (CNDES) cannot be completely eliminated. The presence of uncertainty in the resistance-related design variables can be reduced. It is not possible to accurately quantify uncertainty in the earthquake, wind, and thermomechanical loadings at present. In general, engineering designs are not risk-free, particularly for the dynamic loading applied in the time domain. The associated risk needs to be estimated and managed appropriately and consistently to make the designs of CNDES more sustainable and resilient.

Quantifications of uncertainty in the earthquake and wind loadings for the reliability estimation of CNDES excited in the time domain are emphasized in this chapter. Thermomechanical loading will be discussed in Chapter 9.

4.2 Uncertainty Quantification in Seismic Loadings Applied in the Time Domain

An enormous amount of damage caused by recent earthquakes all over the world indicates that there is room for improvement in designing more seismic load tolerant structures (Azizsoltani et al. 2018). Considering both analytical and experimental investigations, it can be observed that society accepts the current design practices developed over a period of time with a reservation. However, the confidence level reduces significantly just after observing catastrophic damages caused by a severe earthquake. The severity of the damage potential of an earthquake loading is generally expressed in terms of its magnitude and peak ground acceleration (PGA). However, available information indicates that they are not good predictors of damage potential. To address this knowledge gap, acceleration time histories caused by earthquakes need to be considered particularly for the design of critical structures. Recorded time histories are highly irregular, and it is not possible to predict them with certainty. This may be a major reason why the design acceleration time history at a specific site is rarely specified. Since it is highly unpredictable, the associated uncertainties must be incorporated to make the design more seismic load tolerant. In the following discussions, the phrases "earthquake time history" and "acceleration time history" are used interchangeably.

As discussed earlier, the most sophisticated seismic analysis approach requires that the excitation needs to be applied dynamically in the time domain. One of the major sources of uncertainty in the time domain analysis is the prediction of the design acceleration time history of an earthquake for a particular site. Since it is practically impossible to predict, many design codes and guidelines recommend the use of multiple site- and structure-specific time histories at present. The area is evolving and a thorough understanding of the process involved to satisfy the intent of the requirement is essential. In the United States, the seismic design guidelines and requirements are outlined in ASCE Standard – Minimum Design Loads and Associated Criteria for Buildings and Other Structures, generally refer to as ASCE/SEI 7. As new information becomes available, it is frequently updated indicating the evolving nature of the seismic design procedures. ASCE/SEI 7–10 is the basis for all the discussions made in this book.

The primary objective of this section is to generate multiple site-specific design earthquake acceleration time histories fitting the *target probabilistic ground motion response spectrum* (TPGRS) as suggested in ASCE/SEI 7-10. The basic intent is to estimate the probability of failure of a structure when excited by the design earthquake time histories in the presence of major sources of nonlinearity and uncertainty, as intended in the design guidelines.

4.2.1 Background Information

Selection of a suite of design earthquake time histories appropriate for a specific site and structure is a major challenge to engineers. The seismic design maps suggested in design guidelines before 1997 such as Uniform Building Code (UBC) (1997) and National Earthquake Hazards Reduction Program (NEHRP) (FEMA 222A; FEMA 223A) were based on the ground motions with a 10% probability of exceedance (PE) in 50 years. In newer codes, (IBC 2000; FEMA 302; FEMA 303; ASCE/SEI 7-10 2010), PE is considered to be 2% in 50 years, a significant increase in the severity of the design earthquake. The stricter requirement of the design ground motion reflects the changes in the performance objectives of the society and the engineering profession; moving away from life safety (LS) to collapse prevention (CP) to address economic losses (FEMA-P695 2009). These recent guidelines were developed using the risk-targeted maximum considered earthquake (MCE_R) ground motion parameters maps. In ASCE 7-10 (2010) guidelines, seismic design criteria are suggested in Chapter 11. In Section 11.4.7, site-specific ground motion procedures are presented. In Chapter 21 of the same document, site-specific ground motion procedures for seismic design are outlined. In Section 21.2, risk-targeted MCE_R ground motion hazard analysis procedures required to satisfy the intent of the code are clearly stated. These procedures are complicated, can depend on interpretation, and need to be documented for the approval of a design by the appropriate authorities. For routine applications in everyday practice, it may not be easy to satisfy these requirements.

The basic intent of the code is to develop the TPGRS for a specific site so that multiple design earthquake acceleration time histories can be generated by fitting it. To satisfy this requirement, the required TPGRS can be developed in the following way. The ordinates of TPGRS can be generated using Method 1 as outlined in Section 21.2.1.1 of ASCE/SEI 7-10 (2010). They can be obtained by taking the product of the risk coefficients, C_R with the uniform hazard response spectrum (UHRS) (Loth and Baker 2015). UHRS is defined as a 5% damped acceleration response spectrum with a 2% PE within a 50-year period. It corresponds to the annual frequency of exceedance of 0.000404 $\left(1 - \sqrt[50]{1 - 2\%} = 0.000404\right)$. The return period of the ground motion can be estimated as 2475 ($1/0.000404 = 2475$) years. UHRS can be developed by using the programmatically accessible United States Geological Survey (USGS) database.

C_R assures the uniformity of PC of structures using UHRS. It is a period dependent scale factor (SF) and can be estimated following procedures suggested in ASCE 7-10. C_R values are given in figures 22-17 and 22-18 of ASCE 7-10 for periods 0.2 seconds and 1.0 seconds, respectively. For other periods, it can be estimated following the suggested procedures. If the spectral response period is less than or equal to 0.2 seconds, C_R can be taken as the value of C_R at 0.2 seconds. If the spectral response period is greater than or equal to 1.0 seconds, C_R can be taken as the value of C_R at 1.0 seconds. For the spectral response period greater than 0.2 seconds and less than 1.0 seconds, C_R needs to be estimated using linear interpolation.

With the availability of the site-specific UHRS and structure-specific C_R values, the information on the site-specific TPGRS will be available at this stage. It is now necessary to generate a suite of ground motions in the form of acceleration time histories fitting a specific TPGRS (Beyer and Bommer 2007). Several methods are reported in the literature to develop such time histories (Watson-Lamprey and Abrahamson 2006; Jayaram et al. 2011). Two of them are considered. In Alternative 1, a suite of earthquake acceleration time histories can be generated by fitting the required TPGRS with the scaled past recorded data available at the Pacific Earthquake Engineering Research Center (PEER). In Alternative 2, using the information on the seismic activities of the region of interest, the time histories can also be generated numerically using the broadband platform (BBP) available at the Southern California Earthquake Center (SCEC). A response spectrum is developed over a wide frequency range and a fitting strategy should be used to cover this frequency range. It is important now to discuss strategies used for selecting and scaling ground motions to fit a TPGRS for both alternatives.

4.3 Selection of a Suite of Acceleration Time Histories Using PEER Database – Alternative 1

The major task here is to select a suite of ground motions by scaling acceleration time histories available at the PEER database to fit a site-specific TPGRS. The required TPGRS can be generated in two steps as discussed below.

Step 1: In generating TPGRS, the first major task is to generate a UHRS. It can be generated, by using the site-specific hazard curves approach. United States Geological Society (USGS) developed web application-based hazard curves for different locations in the United States. Hazard curves are plots of the annual frequency of exceedance versus the ground motion acceleration for the different periods of a structure. A set of such curves for different periods for Los Angeles are shown in Figure 4.1 for Soil Class D. To generate the required UHRS, a horizontal line corresponding to 2% PE in 50 years is drawn for the various hazard curves. The intersection of the line with the hazard curves for a specific structural period will provide information on the spectral acceleration, as shown in Figure 4.2a. UHRS is a plot of spectral acceleration versus the structural period for specific soil classification.

Site soil conditions are classified into six categories in the United States (ASCE 7-16 2016). They are denoted as Site Classes A to F. Site Class A represents hard rock with measured shear wave velocity greater than 1500 m/s. Site Class B represents rock with shear wave velocity between 760 and 1500 m/s. Site Class F requires site-specific evaluations. FEMA guidelines also suggest that when the soil properties are not known in sufficient detail, Site Class D shall be used.

Step 2: The UHRS generated in Step 1 needs to be multiplied by C_R to obtain a specific TPGRS of interest. The estimation of a structure-specific C_R value is discussed earlier.

Based on the discussions made, response spectra for TPGRS, UHRS, and MCE_R for Los Angeles, New Madrid Seismic Zone, and Charleston are plotted in Figure 4.2a–c,

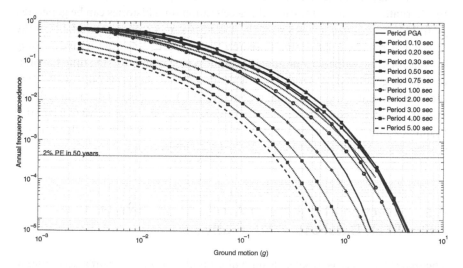

Figure 4.1 Hazard curves for a site located in Los Angeles with coordinates of 118.243° W and 34.054° N, and Site Class D category soil with an average shear velocity of 259 m/s at the depth of 30 m.

respectively. For the site in Los Angeles, since C_R value is equal to 1.0, TPGRS and UHRS spectra are the same. For the other two sites, C_R values are less than 1.0 and the ordinates of TPGRSs are less than UHRS. From the locations of the three curves shown in the figure, no definite conclusion can be made about their characteristics. They depend on the site classification and location. TPGRS for the three sites is

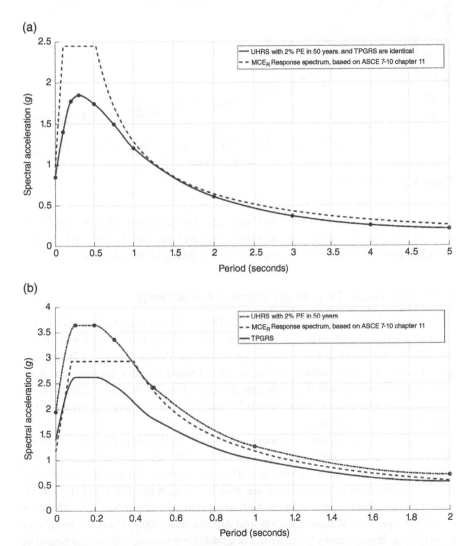

Figure 4.2 Response spectra for TPGRS, UHRS, and MCE$_R$ for the three locations. (a) Los Angeles with Site Class D category soil with an average shear velocity of 259 m/s for the upper 30 m of the site, (b) New Madrid Seismic Zone with Site Class B category soil with an average shear velocity of 1150 m/s for the upper 30 m of the site, (c) Charleston with Site Class D category soil with an average shear velocity of 1150 m/s for the upper 30 m of the site.

(c)

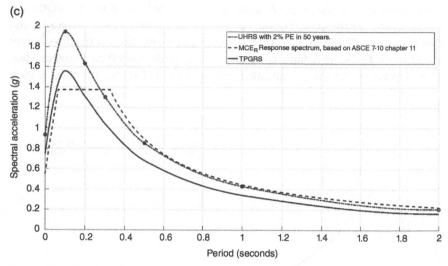

Figure 4.2 (Continued)

shown in a solid line in the figure. In Figure 4.2c, the ordinates of TPGRS are found to be higher than the MCE_R response spectrum over a certain frequency range for the Charleston area indicating complexities in defining all the variables.

4.3.1 Earthquake Time History Selection Methodology

With the availability of the site-specific TPGRS, it is now necessary to generate a suite of earthquake ground motion time histories fitting it. The total number of ground motions to be considered can be different. ASCE7-10 (ASCE 2010) suggested using at least 7. Zimmerman et al. (2015) recommended the use of at least 11 such motions. The algorithm discussed later is capable of generating as many time histories as required. However, results using 11 time histories are presented in the subsequent discussions.

The entire PEER database is considered for selecting the suite of ground motions. Since the ground motions are selected for analyzing two-dimensional structures, the component producing the maximum horizontal response is considered. Initially, the ground motions in the database are scaled to match the TPGRS at the fundamental period of the structure. In the ground motion selection process, the SF closer to unity is desirable (Watson-Lamprey and Abrahamson 2006). However, an SF can be very large in some cases. To select the desirable time histories, the upper bound of SF is suggested to be between 2 and 4 (Bommer and Acevedo 2004). Ground motions with SF more than 4 are eliminated from further consideration.

To select the most appropriate site-specific time histories over a period range, it is necessary to rank the potential candidates in terms of their suitability. To study the suitability factor, the range of the period is considered to be 0.2 and 2 times the fundamental period of the structure under consideration. This range is suggested to be between 0.2 and 1.5 in ASCE 7-10, section 16.1.3.1. The range is then subdivided into 50 equally spaced intervals in the log scale. Information on the lower bound is used to match the ground motions period of the higher vibration modes and the upper bound is used to match the ground motions for highly nonlinear phases. At each of these periods, the differences between the ground motion spectral acceleration spectrum of time history and TPGRS in log scale are estimated.

The total error considering all 50 intervals, generally denoted as the square root of the sum of the squares errors (SRSSE), is estimated for each of the scaled ground motions (Jayaram et al. 2011). It can be represented as:

$$\text{SRSSE} = \sqrt{\sum_{i=1}^{50} \left[\ln S_a(T_i) - \ln S_{a_{\text{TPGRS}}}(T_i) \right]^2} \tag{4.1}$$

where $S_a(T_i)$ and $S_{a_{\text{TPGRS}}}(T_i)$ are the ground motion spectral acceleration and TPGRS, respectively, at period T_i. Eleven ground motion time histories with lower SRSSEs are considered further for the reliability analysis, as will be discussed next.

4.4 Demonstration of the Selection of a Suite of Ground Motion Time Histories – Alternative 1

The earthquake acceleration time history selection process can be demonstrated further with the help of three structures. They are 3-, 9-, and 20-story buildings located in Los Angeles with coordinates of 118.243° W and 34.054° N with Site Class D category soil with a shear velocity between about 180 and 360 m/s for the upper 30 m of the site. The fundamental periods of these structures are 0.87, 2.18, and 3.43 seconds, respectively. Eleven acceleration time histories for these three structures are selected from the PEER database as discussed earlier. The identification nomenclatures used by PEER and the scale factors required to fit TPGRS are summarized in Table 4.1.

Response spectra for the three structures are plotted in Figure 4.3a–c. The spectral accelerations of the selected 11 ground motions and the required TPGRS are shown in the figures. As mentioned earlier, all 11 time histories matched the respective fundamental periods of the three structures. The mean values of the spectral accelerations of the 11 selected ground motions at a specific period are plotted in Figure 4.4a–c. The figures show a good match between the TPGRS and the mean values of the 11 earthquake time histories. To study the uncertainty about the mean, the COV at each period is shown in Figure 4.5a–c.

Table 4.1 Selected time histories for 3-, 9-, and 20-story structures.

Record	3-story		9-story		20-story	
	Record name	Scale factor	Record name	Scale factor	Record name	Scale factor
1	CHICHI/TCU055-N.at2	3.93	CHICHI/TCU089-E.at2	3.38	CHICHI/ILA013-W.at2	3.07
2	NORTHR/RRS318.at2	1.54	CHICHI/ILA013-W.at2	3.18	IMPVALL/H-E07140.at2	2.31
3	SUPERST/B-IVW360.at2	2.97	CHICHI/TCU065-N.at2	1.22	SUPERST/B-PTS315.at2	2.48
4	NORTHR/NWH360.at2	1.09	IMPVALL/H-E08140.at2	2.24	WESTMORL/PTS315.at2	3.43
5	CHICHI/TCU049-N.at2	3.03	LOMAP/HSP090.at2	3.61	CHICHI/TCU076-N.at2	2.23
6	NORTHR/LDM334.at2	2.11	LOMAP/HCH180.at2	2.12	CHICHI/TCU071-E.at2	2.38
7	COYOTELK/G03140.at2	3.58	SUPERST/B-POE360.at2	3.18	DUZCE/BOL000.at2	2.39
8	CHICHI/TCU101-N.at2	3.73	LOMAP/SLC270.at2	3.64	DENALI/ps10047.at2	1.06
9	CHICHI/TCU067-N.at2	1.83	ERZIKAN/ERZ-EW.at2	1.59	LOMAP/HCH090.at2	2.73
10	CHICHI/TCU042-N.at2	3.90	CHICHI/TCU054-N.at2	3.28	CAPEMEND/CPM090.at2	3.47
11	NORTHR/STC180.at2	1.63	IMPVALL/I-ELC270.at2	3.10	ERZIKAN/ERZ-EW.at2	1.65

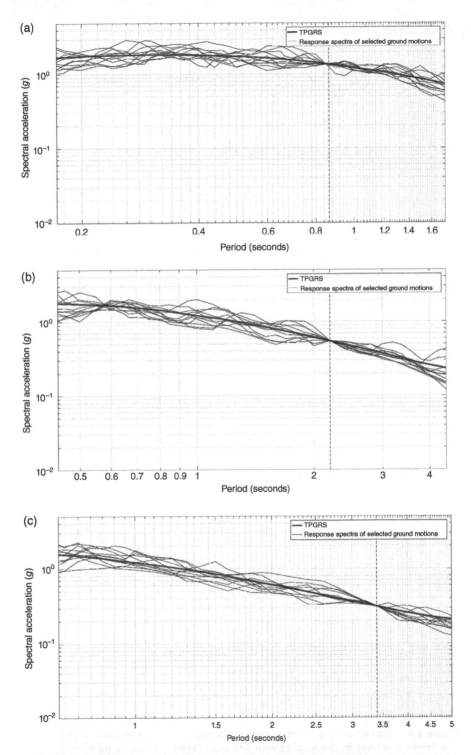

Figure 4.3 Selected ground motions for structures with different fundamental periods in Los Angeles with Site Class D soil. (a) 3-story, (b) 9-story, (c) 20-story.

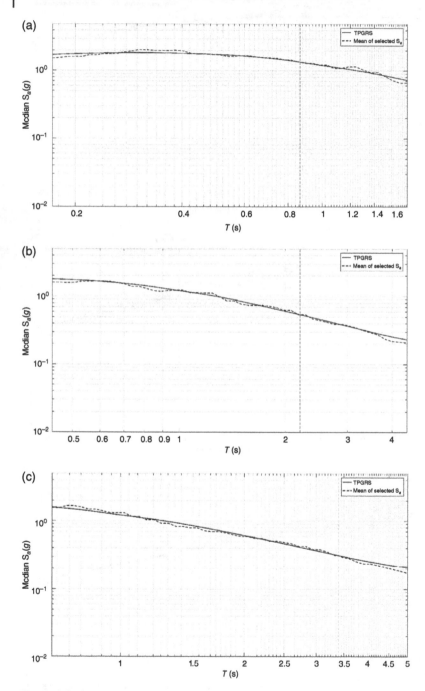

Figure 4.4 Mean value of selected ground motions spectral acceleration for the 3 different structures in Los Angeles with Site Class D soil. (a) 3-story, (b) 9-story, (c) 20-story.

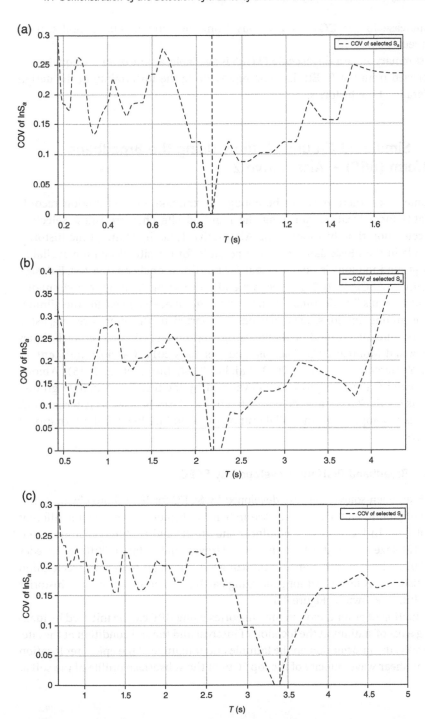

Figure 4.5 COV of selected ground motions' spectral acceleration for three different structures in Los Angeles with Site Class D soil. (a) 3-story, (b) 9-story, (c) 20-story.

An increase in the COV values away from the fundamental period can be observed as expected.

In summary, once a site-specific TPGRS is available, a suite of an earthquake can be selected from the PEER database representing the uncertainty in the design acceleration time histories.

4.5 Simulated Ground Motions Using the Broadband Platform (BBP) – Alternative 2

In some cases, there may not be enough earthquake ground motion records available for a location satisfying the required site characteristics. Selection of acceleration time histories using Alternative 1, i.e. by scaling time histories available in the PEER database is not possible for the site. When the availability of ground motions is not sufficient for nonlinear analyses, the ASCE design guidelines (ASCE 7-10 2010) recommend the use of "appropriate simulated ground motions." Unfortunately, no further guidance is given for the simulation of ground motions. Several studies reported in the literature suggested how to generate ground motions using appropriate simulation techniques (Suárez and Montejo 2005; Cacciola and Deodatis 2011; Cacciola and Zentner 2012; Yamamoto and Baker 2013; Shields 2014; Burks et al. 2015). Among these options, the BBP procedure developed by the SCEC to simulate ground motions (SCEC 2016) appears to be attractive and can be used as Alternative 2. BPP was developed with the collaboration of several national and international partners.

4.5.1 Broadband Platform Developed by SCEC

BBP is an open-source software developed by SCEC for hybrid broadband simulation of ground motions. Several researchers contributed to developing different modules for BBP considering nonlinear site effects and rupture generation concepts (Motazedian and Atkinson 2005; Graves and Pitarka 2010; Schmedes et al. 2010; Mai et al. 2010), and synthesizing low- and high-frequency acceleration time histories. A flow chart for generating earthquake acceleration time histories using BBP is shown in Figure 4.6.

Simulation of ground motion time histories using BBP can be initiated by identifying a list of stations at the location of interest and the soil condition at the site. Generally, the information on the latitude and longitude of the epicenter location and the shear wave velocity of the top 30 m of the subsurface profile of the soil at

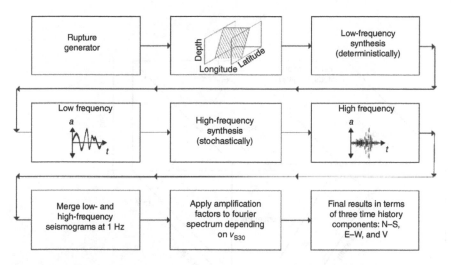

Figure 4.6 Flowchart for generating earthquake ground motions using BBP.

the site, denoted as V_{S30}, is needed. Description of the physical characteristics of the rupture is a fundamental element of BBP. A single-plane fault surface needs to be developed. The information is used by the rupture generator module of BBP. In order to properly generate the rupture, the required parameters are hypocenter (rupture initiation point), fault location, geometry (width, length, dip, and strike), rake (slip direction), and magnitude. These parameters are shown in Figure 4.7.

The next step in the ground motion simulation process is the synthesis of the low- and high-frequency time histories, as shown in the flowchart in Figure 4.6. Low- and high-frequency segments of a time history are simulated separately and then combined to produce a time history of interest (Hartzell et al. 1999). When frequencies are smaller than 1 Hz, a deterministic approach is used. It contains a theoretical representation of wave propagation and fault rupture model to replicate recorded ground-motion amplitudes and waveforms (Graves and Pitarka 2010). When frequencies are greater than 1 Hz, BBP uses a stochastic representation in terms of radiation at the source combining scattering effects and a simplified theoretical representation of wave propagation. Since wave propagation effects and radiation at the source are mainly stochastic at frequencies higher than 1 Hz, different simulation approaches for different frequency bands were used. This is acceptable since there is no uniformity in generating time histories for the ground motions for higher frequencies. At the final step, using information

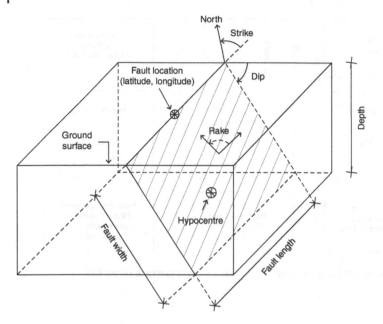

Figure 4.7 Illustration of rupture.

on V_{S30}, an empirical nonlinear site-specific amplification factor is used to the Fourier spectrum of the simulated time history to incorporate local geologic conditions (Walling et al. 2008). The final products are the three acceleration time histories at every station generally denoted as the North–South (N–S), East-West (E–W), and Vertical (V) components.

The BBP version v16.5.0 (SCEC 2016) is used for the simulation of ground motions. As discussed previously, several methods are available in BBP for the simulation of ground motions. The method proposed by Graves and Pitarka (2010), hereafter denoted as the GP method, is utilized for the simulation of ground motions.

4.6 Demonstration of Selection and Validation of a Suite of Ground Motion Time Histories Using BPP

It will be informative to demonstrate how ground motions can be generated using BPP. During the Northridge earthquake of 1994, acceleration time histories were recorded at numerous locations in the Southern California region. Four of them are selected and compared with the time histories generated by BBP at the corresponding locations. These four locations in the Southern California region are (i) Santa Susana, (ii) Alhambra – Fremont, (iii) Littlerock – Brainard, and (iv) Rancho Palos Verdes. Using the BBP v16.5.0 (SCEC 2016) and GP method, the ground motions are simulated at the four locations using the following information reported for the sites. The rupture was generated by the 1994 Northridge earthquake with the magnitude of 6.7. The fault length and width of the fault as shown in Figure 4.7 are 20 and 26 km, respectively. Strike, rake, and dip angles were considered as 122°, 105°, and 40°, respectively, as reported. Response spectra of the measured and simulated versions using BPP of the N-S and E-W components of the ground motions at the four sites are plotted in Figure 4.8. The plots indicate that the measured and simulated ground motions are reasonable. In general, detailed seismic activities of the region are needed to generate time histories of the ground motion using BPP. It may not be the best option if the required information is limited or not available.

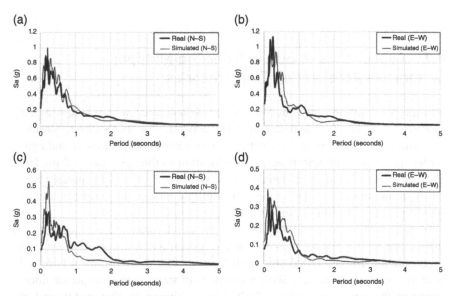

Figure 4.8 1994 Northridge earthquake real and simulated response spectra: (a) Santa Susana N–S, (b) Santa Susana E–W, (c) Alhambra – Fremont N–S, (d) Alhambra – Fremont E–W, (e) Littlerock – Brainard N–S, (f) Littlerock – Brainard E–W, (g) Rancho Palos Verdes N–S, and (h) Rancho Palos Verdes E–W.

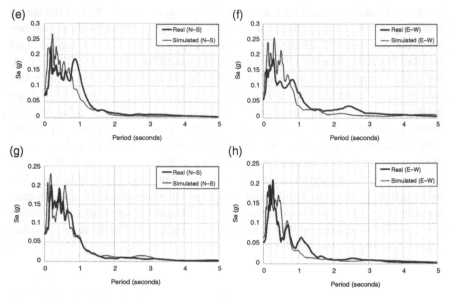

Figure 4.8 (Continued)

4.7 Applications of BBP in Selecting Multiple Earthquake Acceleration Time Histories

Multiple acceleration time histories can also be generated at a specific site using BBP v16.5.0 (SCEC 2016) as discussed in the previous section. However, considerable uncertainty is expected in selecting the locations of the epicenter and hypocenter. To incorporate uncertainty in the location of the epicenter and the fault line, several stations are considered as shown in Figure 4.9. To incorporate uncertainty in the location of the hypocenter, multiple hypocenter locations are also considered, as shown in Figure 4.10.

To demonstrate the generation of multiple acceleration time histories using BBP, a site in the Los Angeles area where necessary information is available is considered. The Raymond Hill Fault located close to Los Angeles is considered for this demonstration. Sixteen locations of stations to incorporate uncertainty in the epicenter and the fault line, as shown in Figure 4.9, and six locations of the hypocenter to incorporate uncertainty in its location, as shown in Figure 4.10, are considered. An earthquake of magnitude 6.7 as reported by USGS (2008) is considered. The length and width of the fault, as shown in Figure 4.7, are considered to be 26 km and 20 km, respectively (Wells and

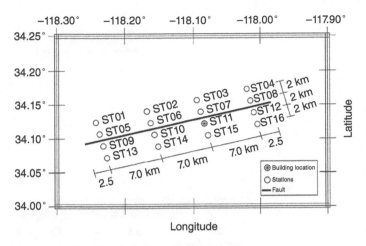

Figure 4.9 Stations, building, and fault location.

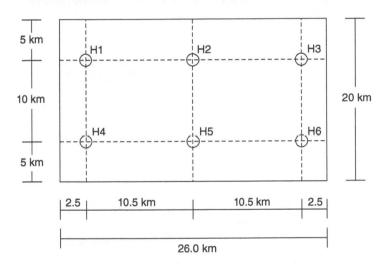

Figure 4.10 Location of hypocenters.

Coppersmith 1994). Strike, rake, and dip are considered to be 75°, 0°, and 90°, respectively.

Considering six hypocenters, 16 stations, and two components (N–S and E–W) per station, a total of 192 acceleration time histories were generated using BPP. The corresponding response spectra of the simulated ground motions are plotted in Figure 4.11. In line with the discussions earlier for Alternative 1, the task is to select the most appropriate 11 acceleration time histories out of a total of 192.

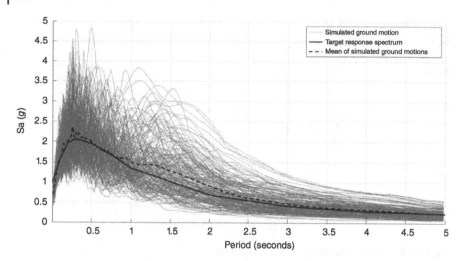

Figure 4.11 Simulated ground motions response spectra and target response spectrum.

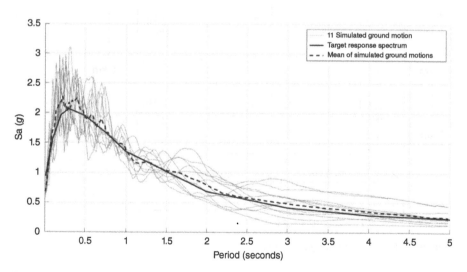

Figure 4.12 Response spectra of 11 selected ground motions and target response spectrum.

The TPGRS as suggested in ASCE/SEI 7-10 is generated as discussed for Alternative 1 in Section 4.2. The 3-story building with a fundamental period of 0.87 seconds is considered. Eleven acceleration time histories with the smallest square root of the SRSSE are selected. They are shown in Figure 4.12. Their mean and TPGRS are also plotted in the figure. They match reasonably well indicating the validity of the BPP approach.

4.8 Summary of Generating Multiple Earthquake Time Histories Using BPP

Multiple acceleration time histories can also be generated using BBP. However, detailed information on the seismic activities of the region must be available. Considering the generation of multiple acceleration time histories using Alternatives 1 and 2, Alternative 1 appears to be more suitable for global applications. Alternative 1 can be implemented with fewer information. Alternative 2 will be appropriate for the sites in highly seismically active regions in the United States. In Chapter 7, another alternative will be suggested to generate multiple earthquake time histories where the use of Alternatives 1 and 2 will be inappropriate.

4.9 Uncertainty Quantification of Wind-Induced Wave Loadings Applied in the Time Domain

4.9.1 Introductory Comments

Offshore structures (OFS) are being built in increasing numbers to satisfy the needs for energy. They are generally submerged in water, away from the coastline, and do not attract much attention from the public. They are expected to be built following the same standards used to build onshore structures (ONS) – buildings, bridges, etc. Unlike OFS, the structural configurations of ONS are visible. In any case, procedures used to analyze and design OFS and ONS appear to be similar, particularly when they are made of steel and subjected only to static loading conditions. The structural arrangements are similar for both ONS and OFS. They are made of structural elements of different sizes and shapes, and connected to each other to satisfy the needs for which they were built. Commonly used resistance-related RVs are related to the geometric shapes and the properties of steel used to build the elements. They include the cross-sectional area and moment of inertia of the shapes. The material properties include the modulus of elasticity, yield stress, strain-hardening ratio, etc. Considering the types of steel structural elements commonly used, the statistical characteristics of uncertainty in resistance-related variables for ONS and OFS structures are expected to be similar.

However, ONS and OFS structures behave very differently under the dynamic loading condition; ONS vibrate in air and OFS vibrate in water. Even if one attempts to model air and water as viscous fluids, the vibration characteristics of ONS and OFS are expected to be different. Dynamic properties are expected to be very different if the structure is vibrating in air and partially submerged in water. Estimating the dynamic responses of them is also expected to be different considering the frequency contents of the seismic and wave loadings and the dynamic properties of the structure. In addition, selections of the design seismic and wave loadings are very difficult, error-prone, and full of uncertainty. Since engineers are not familiar with modeling the wave loading applied in the time domain, quantifying uncertainties in it can be very challenging, even more than quantifying uncertainty in earthquake loading.

At present, at least in the United States, most ONS are designed explicitly to satisfy specified underlying risks, although the risk information is unknown to most practicing engineers. It is essential that OFS are also designed according to the similar risk-based criteria used for ONS. However, a large portion of practicing engineers are not familiar with how to model the wave loading and thus are not expected to know how to quantify uncertainty in it.

There are several types of OFS. However, since they are submerged, structural configurations below the water are often not visible to the public. Two OFSs are shown in Figures 4.13 and 4.14 with appropriate permissions. In Figure 4.13, one type frequently used is referred to as the jacket-type offshore platform (JTOP) to extract oil at the shallow depth of water. They are relatively simpler to design and manufacture. A typical JTOP is a steel frame structure made with tubular shape members that extends from the mud line to above the mean water level. At the top, it supports a deck usually modeled as a rigid structure and its mass is considered to be lumped at the deck level. At the bottom, it is secured by tubular piles attached to its legs. Arrangements of structural elements of a typical JTOP are shown in Figure 4.13. In Figure 4.14 – An arrangement of OFS is shown. Several other types of OFS can be seen using internet.

The reliability estimation of a JTOP excited by the wind-induced wave loading or simply wave loading is needed to make sure that the underlying risk is similar to that of ONS. To meet this objective, the wave loading needs to be applied dynamically in the time domain and the uncertainty associated with it needs to be appropriately incorporated into the formulations. Although not required at present, multiple time histories of the design wave loading may also need to be used to incorporate a considerable amount of uncertainty in it, similar to the earthquake

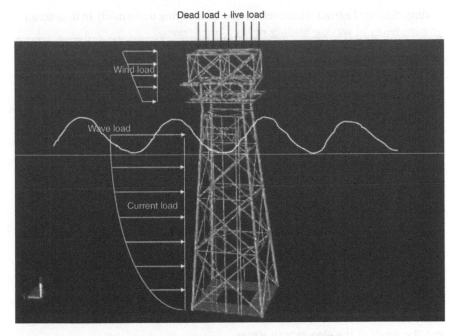

Figure 4.13 A typical jacket-type OFS. *Source:* Henry et al. (2017).

Figure 4.14 An arrangement of OFS. *Source:* https://www.equinor.com/energy/johan-sverdrup.

loading discussed earlier. However, before quantifying uncertainty in it, deterministic modeling of waving loading needs a brief review.

4.9.2 Fundamentals of Wave Loading

The design of OFS, in general, requires a considerable level of expertise. Moreover, the basic analysis and design concepts necessary to study their response behavior when excited by the wave loading are not routinely taught to students in a typical engineering program. Wave loading is very irregular similar to earthquake loading. Analytically representing its basic characteristics can be challenging. A brief review of the deterministic modeling of wave loading is given below.

The governing dynamic equation of motion for OFS and ONS is similar, at least in appearance. Damping is expected to be different since OFS vibrate in water. They are partially submerged in water and the movement of structural elements in water is a major source of energy dissipation that significantly affects their dynamic behavior. The submergence introduces a form of energy dissipation generally known as hydrodynamic damping. In addition, the wave loading exerts forces on all the submerged elements based on their surface areas, wave height, and the depth of the elements in water.

Without losing any generality, the methodology discussed in the following sections is developed for commonly used JTOPs. The governing equation of motion for dynamic structural systems in the matrix notation can be expressed as:

$$\mathbf{M}\ddot{\mathbf{X}} + \mathbf{C}\dot{\mathbf{X}} + \mathbf{K}\mathbf{X} = \mathbf{F} \tag{4.2}$$

where $\mathbf{M}$, $\mathbf{C}$, and $\mathbf{K}$ are the mass, damping, and stiffness matrices, respectively; $\ddot{\mathbf{X}}$, $\dot{\mathbf{X}}$, and $\mathbf{X}$ are the acceleration, velocity, and displacement vectors, respectively; and $\mathbf{F}$ is the external force vector. The important differences in the governing equation between OFS and ONS are the mass matrix $\mathbf{M}$, the damping matrix $\mathbf{C}$, and the external force vector $\mathbf{F}$. In OFS, $\mathbf{M}$ consists of the mass of the structure and the added mass caused by the motion of members in water. The added mass is generally estimated as $\rho(C_m - 1)V$, where ρ is the mass density of water, C_m is the inertia coefficient, and V is the effective volume of the member in water. Modeling of damping also is more complicated. It contains the structural damping and the fluid damping. The fluid damping is caused by the attenuation of movements due to the difference between the velocity of water particles and the structure. The fluid damping is generally estimated as $\rho A U C_d$, where A is the effective projected area of a structural element, U is the velocity of water particles, and C_d is the drag coefficient.

4.9.3 Morison Equation

A conceptual three-dimensional (3D) wave loading is shown in Figure 4.15. It needs to be applied dynamically but only a snapshot of it is shown in the figure. A very complicated computer program was developed to show the complicated dynamic characteristics of wave loading in 3D.

As can be observed in the figure, the modeling of wave loading is not simple. If the information on acceleration and velocity of water is available, using the Morison equation, F_i, the hydrodynamic force per unit length acting normal to the axis of cylindrical member i, can be calculated as the summation of the drag and inertia forces. As suggested by American Petroleum Institute (API 2007), the Morison equation can be defined as:

$$F_i = F_{i,d} + F_{i,I} \qquad (4.3)$$

where $F_{i,d}$ and $F_{i,I}$ are the drag and the inertia forces, respectively, per unit length of the member i. They are evaluated as:

$$F_{i,d} = C_d \rho / 2 A \ U|U| \qquad (4.4)$$

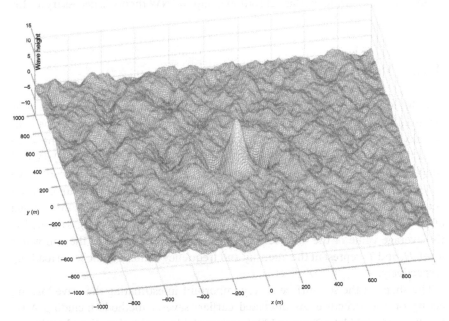

Figure 4.15 A conceptual water surface. *Source:* Vazirizade et al. (2019).

and

$$F_{i,I} = C_m \rho V \frac{\delta U}{\delta t} \tag{4.5}$$

where C_d is the drag coefficient, ρ is the density of water, A is the projected area per unit length, and V is the displaced volume of the cylinder per unit length ($\pi D^2/4$ for circular cylinders), D is the effective diameter of the circular cylindrical member including marine growth, U and $|U|$ are the conjugate components of the velocity vector (due to wave and/or current) of the water normal to the axis of the member, C_m is the inertia coefficient, and $\frac{\delta U}{\delta t}$ is the component of the local acceleration vector of the water normal to the axis of the member.

The evaluation of the Morison equation requires information on the fluctuation of the sea surface about the mean sea level, as a function of time. There are several methods available in the literature for this purpose. Some of the classical deterministic methods for estimating the sea surface fluctuations are Airy, Stocks, and Cnoidal (Dawson 1983). However, none of these methods models the water surface elevation precisely. To address this problem, the New Wave (NW) theory was proposed. It is deterministic in nature and accounts for the spectral content of the sea (Tromans et al. 1991). A brief discussion on the generation of the water surface elevation using the NW theory is necessary at this stage.

4.10 Modeling of Wave Loading

For the accurate estimation of the deterministic response behavior OFS, it is essential to incorporate wave loading in the dynamic governing equation as realistically as possible. Two important items need to be considered in modeling wave loading: (i) the wave height representing the intensity of the wave; and (ii) the real shape of the wave profile. The wave height is represented by the significant wave height, H_s and the wave profile can be described by the mean zero-up crossing period, T_z. H_s and T_z are two important meteoceanic parameters. H_s is the average of the upper third of the wave heights. The zero-up crossing period, T_z, is defined as the average value of the time between successive up crossings of the still water level. H_s and T_z represent the intensity and frequency content of the wave loading, respectively.

The shape of the sea surface is very important in estimating the wave loading acting on the structure. As discussed earlier, several methods including Airy, Stocks, and Cnoidal (Dawson 1983) are available to simulate the elevation of the sea surface but since they fail to consider the uncertainty in the sea waves, other options need to be explored.

4.10.1 Wave Modeling Using the New Wave Theory

The NW theory analytically simulates the most probable shape of the sea surface by a Gaussian random process during an extreme event. For the nonlinear dynamic analysis, the load history is also very important in modeling the sea surface and the information on it needs to be incorporated into the formulation. Modeling the time domain sea surface over a long period of time can be difficult and cumbersome. The problem can be overcome by using the NW theory. It simulates many hours of the time domain simulation of wave loading in a more computationally efficient way (Cassidy 1999; Cassidy et al. 2001, 2002; Tromans et al. 1991). Using the NW theory, the surface elevation, η, as a function of time can be expressed as:

$$\eta(x, t) = \frac{\alpha}{\sigma^2} \sum_{n=1}^{N} [S(\omega_n)d\omega] \cos(k_n x - \omega_n t) \tag{4.6}$$

where x is the location and $x = 0$ represents the wave crest, t is the time relative to the initial position of the crest, k_n is the wave number of the nth component, ω_n is the angular frequency of the nth component, α is the NW crest elevation, $S_{\eta\eta}(\omega_n)$ $d\omega$ is the surface elevation spectrum, and σ^2 is the variance corresponding to that wave spectrum, i.e. the area under the spectrum. The most commonly used surface elevation spectrum is Jonswap (Hasselmann et al. 1973). It is mathematically expressed as:

$$S(\omega) = \bar{\alpha}g^2\omega^{-5}e^{-\beta_j(\omega/\omega_p)^{-4}}\gamma^a \tag{4.7}$$

where γ is the peakedness parameter (it is 3.3 for Jonswap and 1 for Pearson-Moskowitz), ω is the angular frequency, ω_p is the peak frequency; the frequency at which the spectrum peaks, β_j is the shape factor and equals to 1.25, g is the gravitational acceleration, and a and $\bar{\alpha}$ are modified Phillips constants. They can be estimated as:

$$a = e^{-\dfrac{(\omega - \omega_p)^2}{(2\sigma^2\omega_p^2)}} = e^{-\left(\dfrac{\omega/\omega_p - 1}{\sigma\sqrt{2}}\right)^2} \tag{4.8}$$

and

$$\bar{\alpha} = 5.058 \left(\frac{H_s}{T_p^2}\right)^2 (1 - 0.287 \ln \gamma) \tag{4.9}$$

where H_s and T_p are the significant wave height and the peak frequency (the period associated with the peak of the spectrum), respectively. Denoting T_z as the mean zero-crossing period, T_p can be calculated as (Cassidy 1999):

$$T_p = T_z\left(0.327\,e^{-0.315\gamma\,+\,1.17}\right) \tag{4.10}$$

where γ is defined earlier, H_s and T_z are site-specific and depend on the location of the structure.

4.10.2 Wheeler Stretching Effect

The motion of the water particles and the forces they generate on the structure are a function of depth z, as shown in Figure 4.16. It can be calculated as (Journée and Massie 2001):

$$F_n(z) = \frac{\cosh\left[k_n(d + z)\right]}{\sinh\left(k_n d\right)} \tag{4.11}$$

where F_n is called the attenuation factor with respect to sea level (Y-axis) for the nth component. The velocity of the water particles in the Z-axis as a function of time (X-axis) and depth is shown in Figure 4.16. In order to make these values valid above the mean water-level, $F_n(z)$ in Eq. 4.11 needs to be modified by either Wheeler stretching or Delta stretching interpolation (Wheeler 1970). Considering Wheeler stretching, it can be mathematically represented as:

$$F_n(z) = \frac{\cosh\left[\dfrac{k_n(d + z)}{1 + \eta/d}\right]}{\sinh\left(k_n d\right)} \tag{4.12}$$

where d is the depth under consideration, η is the free water surface, and z is the variable that can change from the seabed to η. By using the Wheeler stretching method, the velocity of water particles in different depths can be calculated as shown in Figure 4.16.

4.10.3 Three-Dimensional Directionality

As can be observed in Figure 4.15, the wave loadings are expected to be different in the two horizontal directions: essentially a function of θ, the direction of the wave. In order to address the directionality-related issue, the directional spectrum concept can be used. It can be mathematically represented as:

$$S(\omega, \theta) = S(\omega)\,D(\theta) \tag{4.13}$$

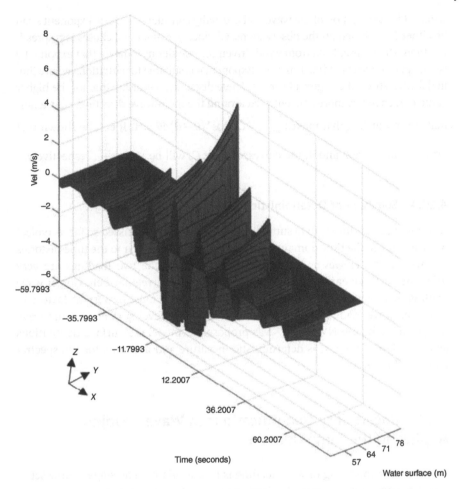

Figure 4.16 Particle velocity as a function of depth and time based on water surface level. *Source:* Vazirizade et al. (2019).

where $S(\omega)$ *is the surface elevation spectrum and* $D(\theta)$ *is the directional spreading* function. Many different functions were reported in the literature to consider the directional spreading function (Soares 1998). The aim was to generate higher weights for the directions closer to the main direction. According to API (2007), $D(\theta)$ is represented by a trigonometric function as:

$$D(\theta) = C_s \cos^s\left(\theta - \overline{\theta}\right) \tag{4.14}$$

where $\bar{\theta}$ is the direction of the wave and is usually considered as zero, exponent s can be either 2 or 4 based on the observations. Obviously, exponent 4 causes less spreading than when it is 2. Data from wind-driven sea conditions indicate that exponent 2 is the appropriate spreading function. Exponent 4 is suitable for situations where limited fetch restricts the degree of spread. These discussions indicate that for the higher value of s, energy is more concentrated around the main wave direction. C_s is a normalizing coefficient that makes $\int_{\bar{\theta}-\pi/2}^{\bar{\theta}+\pi/2} C_s \cos^s\left(\theta - \bar{\theta}\right)d\theta = 1$. It can be shown that when s values are 2 and 4, the corresponding C_s will be $\dfrac{\pi}{2}$ and $\dfrac{3\pi}{8}$, respectively.

4.10.4 Summary of Deterministic Modeling of Wave Loading

It is important at this stage to summarize what has been discussed so far. A typical wave loading in the time domain is extremely irregular similar to the time histories of earthquakes. However, the frequency contents of the two loadings are very different. Just like earthquake loading, capturing the irregular behavior of wave loading mathematically can be very challenging. The most important factors in modeling wave loading are the profile of the water level surface and the wave height. Initially, the NW theory was proposed to model sea surface fluctuations as a function of time. It is deterministic in nature and accounts for the spectral content of the sea.

4.11 Uncertainty Quantifications in Wave Loading Applied in the Time Domain

Deterministic modeling of wave loading is now available. The stage is now set to quantify uncertainty in wave loading applied in the time domain.

4.11.1 Uncertainty Quantification in Wave Loading – Three-Dimensional Constrained New Wave (3D CNW) Concept

The NW concept discussed in the previous section does not consider the randomness of the sea waves. The constrained new wave (CNW) concept can be used to consider randomness in wave loading (Mirzadeh 2015). Using the CNW theory, many hours of random wave loading in the time domain can be simulated in a more computationally efficient manner. The NW concept provides a deterministic wave profile for a predetermined height accounting for the spectral composition of the sea and the CNW concept adds uncertainty to the formulation. Mathematically, the basic CNW concept can be expressed as:

$$
\eta_c(x, t) = \eta_r(t) + r(x, t) \left[\alpha - \sum_{n=1}^{N} a_n \right] + \left(\frac{-\dot{r}(x, t)}{\lambda^2} \right) \left[\dot{\alpha} - \sum_{n=1}^{N} \omega_n b_n \right]
$$

(4.15)

where $\eta_r(t)$ is the random surface elevation above the mean water level; $r(x, t)$ is the unit NW, which is the normalized water surface level; α is a predetermined constrained amplitude of the wave; $\sum_{n=1}^{N} a_n$ represents the random surface elevation at $t = 0$ [or $\eta_r(0)$]; $-\dot{r}(x, t)$ is the slope of the unit NW; λ^2 is obtained from the second spectral moment of the wave energy spectrum ($\lambda^2 \sigma^2$); $\dot{\alpha}$ is the predetermined constrained slope for a crest, $\dot{\alpha} = 0$; and $\sum_{n=1}^{N} \omega_n b_n$ is the random surface slope at $t = 0$ or $\dot{\eta}_r(0)$.

The random surface elevation above the mean water level $\eta_r(t)$ in Eq. 4.15 for a uni-directional wave can be represented as (Mirzadeh 2015):

$$
\eta_r(t) = \sum_{n=1}^{N} (a_n \cos(\omega_n t) + b_n \sin(\omega_n t))
$$

(4.16)

where ω_n is the angular frequency of the nth component as discussed earlier, $a_n = rn_{a_n} \sqrt{S(\omega_n) d\omega}$ and $b_n = rn_{b_n} \sqrt{S(\omega_n) d\omega}$; a_n and b_n are Fourier components, which are Gaussian RVs with zero mean and a standard deviation related to the wave energy spectrum at the corresponding discrete frequency which is $\sigma_n = \sqrt{S(\omega_n) d\omega}$.

Two sample water surfaces generated by the NW and CNW theories are compared in Figure 4.17. To generate these figures, H_s, T_z, and α are considered to be 12 m, 10 seconds, and 15 m, respectively. Consequently, the maximum elevation of the crest, which occurs at $t = 0$ and $x = 0$, is the same. However, it can be observed that CNW adds uncertainty to NW. It is to be noted that the if two water surface profiles are generated using NW and CNW for the same set of parameters, they will share the same maximum elevation of the crest, but the profiles will be different indicating the presence of uncertainty.

To clarify the CNW theory further, the three components of CNW denoted hereafter as CNW1, CNW2, and CNW3, corresponding to the terms $\eta_r(t)$, $r(x, t) \left[\alpha - \sum_{n=1}^{N} a_n \right]$, and $\left(\frac{-\dot{r}(x, t)}{\lambda^2} \right) \left[\dot{\alpha} - \sum_{n=1}^{N} \omega_n b_n \right]$, respectively, are considered separately and plotted in Figure 4.18. CNW2 represents the scaled water surface

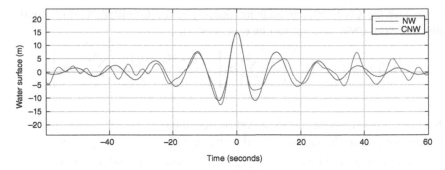

Figure 4.17 Wave profile of NW and CNW concepts. *Source:* Vazirizade et al. (2019).

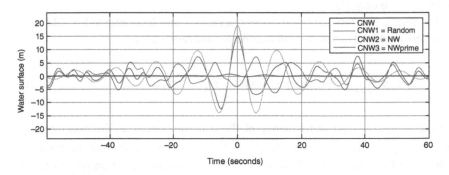

Figure 4.18 CNW and its three components. *Source:* Vazirizade et al. (2019).

elevation according to NW. The plots clearly indicate the uncertainty in NW. With the CNW theory, the water surface can be simulated in the time domain. This approach is an alternative to widely used regular wave theories, such as Airy and Stokes' 5th-order wave (Cassidy 1999; Cassidy et al. 2001, 2002; Tromans et al. 1991).

4.11.2 Three-Dimensional Constrained New Wave (3D CNW) Concept

In order to convert CNW from two-dimensional (2D) to there-dimensional (3D), the directionality effect needs to be considered. CNW generates water surface elevation considering uncertainty in the shape of the waves including the wave height and the frequency content. However, it does not include the directionality effects. Incorporating directionality information in 2D CNW will result in 3D

CNW (Mirzadeh et al. 2016). Essentially, directionality considers the randomness in the two horizontal directions by adding another direction along the Y axis. In 2D, the wave moves in the X direction, and as discussed before, the crest height changes over time in the X direction. However, in 3D, the crest height also changes in the perpendicular direction of the wave motion in the Y axis. To include the effects of wave directionality, θ is introduced as explained in Eqs. 4.13 and 4.14, which effect the values of $a_{m,n}$ and $b_{m,n}$, as discussed next. Thus, 3D CNW can be defined as:

$$
\begin{aligned}
\eta_c(x,y,t) = \eta_r(t) + r(x,y,t)\left[\alpha - \sum_{m=1}^{M}\sum_{n=1}^{N}a_{m,n}\right] \\
+ \left(\frac{-\dot{r}(x,y,t)}{\lambda^2}\right)\left[\dot{\alpha} - \sum_{m=1}^{M}\sum_{n=1}^{N}\omega_n b_{m,n}\right]
\end{aligned}
\tag{4.17}
$$

where $r(x,y,t)$ is the unit NW in the X and Y directions, $\alpha, \dot{\alpha}$, and λ^2 are defined earlier for 2D CNW, $\sum_{m=1}^{M}\sum_{n=1}^{N}a_{m,n}$ is the random surface elevation at $t = 0$ (or $\eta_r(0)$), $-\dot{r}(x, y, t)$ is the slope of the unit NW obtained from the second spectral moment of the wave energy spectrum $(m_2 = \lambda^2\sigma^2)$, $\dot{\alpha}$ represents the predetermined constrained slope; for a crest, $\dot{\alpha} = 0$, $\sum_{m=1}^{M}\sum_{n=1}^{N}\omega_n b_{m,n}$ is the random surface slope at $t = 0$ or $\dot{\eta}_r(0)$, M is the number of wave directions, N is the number of frequency components, and $\eta_r(t)$ is defined later. This 3D CNW can be generated by a finite number $(M \times N)$ of sinusoidal wave components. This is the most comprehensive equation, which includes directionality, randomness, and frequency content. For $M = 1$, it will be 2D instead of 3D waves. The random surface elevation for the 3D waves as a function of t can be defined as follows:

$$
\eta_r(t) = \sum_{m=1}^{M}\sum_{n=1}^{N}(a_{m,n}\cos(\omega_n t) + b_{m,n}\sin(\omega_n t))
\tag{4.18}
$$

where $a_{m,n} = rn_{a_n}\sqrt{S(\omega_n, \theta_m)d\omega\, d\theta}$ and $b_{m,n} = rn_{b_n}\sqrt{S(\omega_n, \theta_m)d\omega\, d\theta}$.

The 3D effect of water surface level shown in Figure 4.19 can now be represented mathematically with the help of the proposed 3D CNW concept. Suppose two legs of a JTOP are separated by a distance. They are expected to be subjected to two different wave profiles as shown in Figure 4.19. The figure demonstrates the difference between water surface elevations at the upwave and the downwave legs of a JTOP. In this figure, the parameters are $H_s = 12$ m and $T_z = 10$ s. Furthermore, the differences between NW and CNW wave profiles can be observed by comparing Figures 4.19a, b. In CNW, the shape of the wave profiles

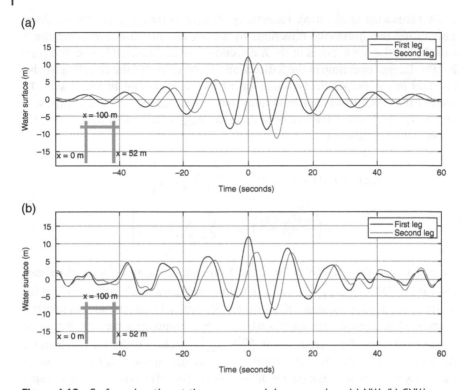

Figure 4.19 Surface elevation at the upwave and downwave legs (a) NW; (b) CNW.

changes over the time; therefore, the wave produces different forces at upwave and downwave legs.

4.11.3 Uncertainty in the Wave Height Estimation

To estimate the uncertainty in the intensity of the wave, a distribution for the wave height, H, in the location of interest is required. It is very common in the profession to estimate the uncertainty in the intensity of the wave height H by the joint distribution of the significant wave height H_s and the zero-up crossing period T_z (Eik and Nygaard 2003; Haver and Winterstein 2008). The information on them is usually recorded in a three-hour duration and both parameters are estimated for this duration. After collecting the data over a period of time, the information is summarized in a tabular form, and graphically presented in the form of a scatter diagram, as discussed in Vazirizade et al. (2019). It represents the total number of observations for each pair of H_s and T_z. The values of H_s and T_z are generally considered at the center of each interval, i.e. when H_s is 1.5 m, it represents the interval of 1 and 2 m. T_z is defined similarly.

4.11.4 Uncertainty Quantification of Wave Loading

With the availability of the data discussed earlier, the major challenge is to find an appropriate mathematical model to represent uncertainty in wave loading. A literature review will indicate the joint probability density function (PDF) of T_z and H_s is generally represented as the multiplication of a conditional distribution for T_z given H_s and the marginal distribution of H_s (Eik and Nygaard 2003; Bitner-Gregersen 2010, 2015). It can be mathematically represented as:

$$f_{T_z,H_s}(t_z, h_s) = f_{T_z|H_s}(t_z|h_s) f_{H_s}(h_s) \qquad (4.19)$$

The joint distribution model represented by Eq. 4.19 is very helpful since it includes both the frequency content and amplitude of the sea waves (Bitner-Gregersen and Haver 1989). The most common distribution used for the conditional PDF (the first term on the right-hand side of Eq. 4.19) is lognormal (Haver 1985). Weibull and Rayleigh distributions are generally used for the marginal distribution of H_s (the second term on the right-hand side of Eq. 4.19) (Nordenström 1973; Shariff and Hafezi 2012). Subsequently, it was reported that lognormal distribution for conditional distribution T_z given H_s and Weibull for the marginal distribution of H_s are the best choices (Bitner-Gregersen and Haver 1989; Mathisen and Bitner-Gregersen 1990). Even though this model was initially developed for the Norwegian Sea, it was shown that this could be adjusted for any location (Bitner-Gregersen et al. 1995). Finally, Sigurdsson and Cramer (1996) recommended lognormal distribution for the conditional distribution and Weibull distribution for the marginal distribution of H_s and showed that these were reliable mathematical models. They are considered here.

The conditional PDF of $f_{T_z|H_s}(t_z|h_s)$ considering it is lognormally distributed can be mathematically expressed as:

$$f_{T_z|H_s}(t_z|h_s) = \frac{1}{\sqrt{2\pi}\,\zeta_{T_z|H_s}(t_z|h_s)} \exp\left[\frac{\left(\ln(t_z|h_s) - \lambda_{T_z|H_s}\right)^2}{2\zeta_{T_z|H_s}^2}\right] \qquad (4.20)$$

where $\lambda_{T_z|H_s}$ and $\zeta_{T_z|H_s}$ are the two parameters of the lognormal distribution. They are essentially mean and standard deviation of the natural logarithm of the original data. For the conditional lognormal distribution, the two parameters can be shown to be $\lambda_{T_z|H_s} = a_0 + a_1 h_s^{a_2}$ and $\zeta_{T_z|H_s} = b_0 + b_1 \exp(b_2 h_s)$ (Sigurdsson and Cramer 1996). Based on the available data and the procedure suggested in Vazirizade et al. (2019), $\lambda_{T_z|H_s}$ and $\zeta_{T_z|H_s}$ can be estimated. Using information on $\lambda_{T_z|H_s}$, the parameters a_0, a_1, and a_2 can be estimated by curve fitting the available data. Similarly, b_0, b_1, and b_2 can be estimated using the information on $\zeta_{T_z|H_s}$.

As discussed earlier, the marginal distribution of H_s is considered to follow the Weibull distribution. The Weibull distribution is a type III extreme value distribution (Haldar and Mahadevan 2000a). The PDF and cumulative distribution function (CDF) of H_s, $f_{H_s}(h_s)$ can be expressed as:

$$f_{H_s}(h_s) = \frac{\beta(h_s - \gamma)^{\beta-1}}{\alpha^\beta} \cdot \exp\left\{ -\left(\frac{h_s - \gamma}{\alpha}\right)^\beta \right\} \tag{4.21}$$

and

$$F_{H_s}(h_s) = 1 - \exp\left\{ -\left(\frac{h_s - \gamma}{\alpha}\right)^\beta \right\} \tag{4.22}$$

where α, β and γ are the shape (slope), scale, and location (shift or threshold) parameters, respectively. α and β are always positive. When $\beta = 1$, Eq. 4.21 represents the exponential distribution, and when $\beta = 2$, it represents the Rayleigh distribution. However, γ can be positive or negative. If it is zero, the three-parameter Weibull distribution will reduce to the two-parameter Weibull distribution. γ is the minimum value for which the CDF is non-zero. According to the definition of H_s which is the average of the upper third of the wave heights, it is always positive and cannot be zero. The joint distribution of H_s and T_z is shown in Figure 4.20 (Vazirizade et al. 2019).

To estimate the uncertainty in the wave height H based on the uncertainty in the significant wave height H_s for a region, Forristall and Cooper (1997) suggested a conditional CDF of H on H_s. It is generally referred to in the literature as the Forristall wave height distribution. It can be mathematically expressed as:

$$F_{H|H_s}(h|h_s) = 1 - \exp\left[-2.26\left(\frac{h}{h_s}\right)^{2.13} \right] \tag{4.23}$$

The unconditional CDF of H can be derived as:

$$F_H(h) = \int_{h_s} F_{H|H_s}(h|h_s) \times f_{H_s}(h_s).dh_s \tag{4.24}$$

Thus, by combining Eqs. 4.21 and 4.23, the uncertainty in H can be obtained.

4.11.5 Quantification of Uncertainty in Wave Loading

By integrating the information provided in Sections 4.9–4.11 and wave activities of the region, multiple time histories of waves can be generated. Then REDSET can

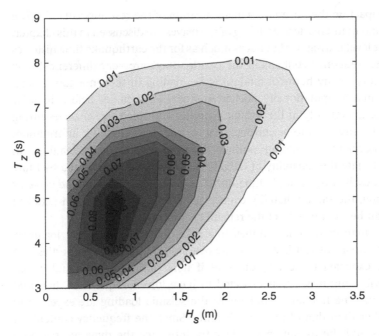

Figure 4.20 Joint distribution of H_s and T_z. *Source:* Vazirizade et al. (2019).

be used to extract reliability information as will be discussed in more detail in Chapters 5 and 7.

4.12 Wave and Seismic Loadings – Comparisons

Even though both wave and earthquake loadings fall into the category of dynamic lateral loading, they act very differently. The wave force is applied to all members, which are below the water surface level. This force is usually higher closer to the water surface and attenuated along the depth. The seismic loading acts at the base of the structure and all the members are excited by the same loading. It can be observed that the frequency contents of wave and seismic loadings are very different and sources of uncertainty in modeling damping need to be considered very carefully. In general, it may not be possible to postulate whether the wave or seismic loading will be critical for the design of an OFS at a particular location. Reliability estimations of a JTOP excited in the time domain by the wave and seismic loadings will be demonstrated and compared in Chapter 7.

The discussions made here clearly indicate the quantifications of uncertainty in the seismic and wave loadings will be very different. Obviously, they will have a significant impact on the uncertainty quantification of the responses of OFS. The uncertainty and mathematical modeling of the waves are discussed in this chapter. The sea water level is irregular but not as much as for the earthquake time histories indicating the frequency contents of the two loadings are very different. Using the CNW theory, many hours of random wave loading in the time domain can be simulated in a computationally efficient manner. As discussed before, in order to address the uncertainty in the seismic loading, multiple time histories fitting the TPGRS at a site need to be considered to extract the reliability information. A similar concept of considering multiple time histories for wave loading can be used to estimate the reliability of OFS. When an OFS is represented by finite elements and excited by multiple time histories generated either by the wave or seismic loading, the same reliability concept REDSET, as discussed in detail in Chapter 5, can be used to extract the reliability information.

Even after quantifying uncertainties in the earthquake and wave loadings, another very important fact just cannot be overlooked. Dynamic properties of a structure are expected to be very different if it is vibrating in air and water. A structure vibrating in water is expected to have lower frequencies than that vibrating in air. The frequency contents of the seismic loading are expected to be much higher than that of the wave loading. Since the frequency contents of the structures and the excitation significantly influence the dynamic response behavior in the presence of different types of damping, the knowledge obtained from the design of ONS may not be applicable to the design of OFS including JTOPs. As discussed earlier, the selection of design seismic ground motions can be very error-prone. Similar challenges can also be faced in modeling the uncertainty in wave loading. The challenges will have a compounding effect on the nonlinear time domain analysis of both ONS and OFS structures.

4.13 Concluding Remarks

Discussions made in this chapter clearly indicate the complexities in quantifying uncertainties in both seismic and wave loadings, particularly when they need to be applied in the time domain. One of the major conclusions is that both loadings must be represented by multiple time histories to consider uncertainty in the frequency contents of the excitations. For generating site- and structure-specific multiple earthquake time histories, information on TPGRS needs to be developed first. Two alternatives – scaling the PEER database and BBP using the information on the detailed seismic activities of the region – can be used to generate the required numbers of time histories. It was discussed that by using the CNW concept, an unlimited number of wave profiles could be generated in the time domain at a particular site. They contain the same meteoceanic properties, but their profiles and frequency contents will be different.

A novel risk evaluation technique denoted as REDSET to estimate the risk of CNDES excited in the time domain by the earthquake and wind loadings discussed in this chapter will be formally presented in Chapter 5. REDSET will be verified for the earthquake and wave loading with the help of numerous examples in Chapters 6–8.

5

Reliability Assessment of Dynamic Systems Excited in Time Domain – REDSET

5.1 Introductory Comments

Discussions in the previous chapters set the stage for formally presenting the novel reliability estimation procedure REDSET. The acronym REDSET stands for *Reliability Evaluation of Dynamic Systems Excited in Time* Domain. It is specifically developed for the reliability estimation of *complex nonlinear dynamic engineering systems* (CNDES). The justifications for developing it were given in Chapter 1. In summary, there is no method currently available to estimate the reliability of CDNES. The underlying risk can be estimated by conducting dozens to a few hundred deterministic nonlinear finite element (FE)-based dynamic analyses excited in the time domain. The basic concept of REDSET now needs to be formally presented.

The state of the art for reliability analysis of engineering systems with explicit limit state functions (LSFs) is very advanced (Haldar and Mahadevan 2000a). No similar statement can be made for reliability assessment with implicit LSFs. Three alternatives for the reliability analysis for this class of problem were identified in Section 3.2. Two of them were eliminated from further consideration. Only one of them using the response surface method (RSM)-based approach remains a possibility. However, RSM was developed for different applications, and the basic concept cannot be used for the reliability analysis of engineering systems for several reasons. A conceptual framework of the approach was presented in Chapter 3 for approximately generating explicit expressions in the failure region for implicit LSFs containing information on structural responses, by extensively modifying the basic RSM concept. The response surfaces generated with this significantly improved procedure were denoted as improved response surfaces (IRSs). They were generated in the form of second-order polynomials by fitting the response data obtained from conducting a few deterministic FE-based nonlinear dynamic

Reliability Evaluation of Dynamic Systems Excited in Time Domain: Alternative to Random Vibration and Simulation, First Edition. Achintya Haldar, Hamoon Azizsoltani, J. Ramon Gaxiola-Camacho, Sayyed Mohsen Vazirizade, and Jungwon Huh.
© 2023 John Wiley & Sons, Inc. Published 2023 by John Wiley & Sons, Inc.

analyses of a structure excited in the time domain, using any available computer program. At least three advanced experimental design schemes were proposed, but the most attractive one consisting of the central composite design (CCD) scheme could not be implemented for the reliability evaluation of CNDES in most cases. The major task at this stage is to make some of the proposed advanced design schemes implementable in the context of REDSET so that the underlying reliability can be estimated efficiently and accurately. The basic idea is that if an implementable IRS can be generated explicitly using these advanced concepts, the corresponding LSF will also be explicit, and then the underlying risk can be easily estimated using FORM.

The major hurdle blocking the development of REDSET at this stage is making the advanced sampling design schemes proposed in Chapter 3 implementable for routine applications in everyday practice. To meet this objective, several advanced mathematical concepts need to be integrated to produce compounding beneficial effects. The necessary tasks need to be executed in the proper sequence to obtain the desired results. If successful, the knowledge gap that exists today for the reliability estimation of CNDES will be eliminated. Since it is not possible to design CNDES completely risk-free, the associated risk needs to be appropriately managed. REDSET is expected to be a major tool for designing more sustainable and resilient CNDES.

To develop REDSET efficiently, the following items/information must be readily available:

1) A computer program capable of conducting time domain nonlinear FE analysis of structures must be available.
2) A list of LSFs of interest and the corresponding performance requirements, as discussed in Chapter 3, must also be available.
3) Statistical information on all random variables (RVs) in the IRSs of interest must be available, in terms of mean, variance, distribution, correlation characteristics, etc., as discussed in Chapter 2. Users need to be familiar with the information on uncertainty associated with all the resistance-related RVs in the formulation. The procedures to quantify uncertainty in the earthquake and wave loadings applied in the time domain are discussed in detail in Chapter 4. The uncertainty associated with the thermomechanical loading will be discussed in Chapter 9. Users need to be familiar with uncertainty quantification in all RVs before moving forward.
4) Users need to be familiar with the basic sampling design schemes to generate an IRS discussed in Chapter 3. A review of the materials is expected to accelerate the learning process.

5.2   A Novel Reliability Estimation Concept – REDSET

The information in the first few chapters, if integrated properly, will produce the first version of REDSET. However, it will not be efficient, accurate, and robust enough to extract reliability information for CNDES. In some cases, the basic concept cannot even be implemented. Several major important improvements to the initial version of REDSET need to be undertaken in a sequential and systematic way to develop REDSET. This is discussed next.

5.2.1   Integration of Finite Element Method, Improved Response Surface Method, and FORM

The first important challenge in the development of REDSET is the generation of an IRS in the failure region explicitly considering all major sources of uncertainty in the design variables. FORM iteratively locates the most probable failure point (MPFP), explicitly considering the statistical information on the uncertainty associated with the design variables. To assure that an IRS will be generated in the failure region explicitly incorporating uncertainty information, it is desirable to integrate the generation process of an IRS with FORM.

To initiate the iterative process of FORM, the coordinates of the first checking point are generally assumed to be the mean values of all the RVs. To facilitate the integration process, the coordinates of the first checking point will remain the same for generating an IRS in the context of REDSET. This will provide the coordinates of the first center point for generating an IRS. Then, using any of the experimental sampling design schemes discussed in Section 3.11, an engineering system represented by FEs will be dynamically excited in the time domain to obtain the maximum value of the response (maximum lateral deflection and inter-story drift, maximum axial load and bending moment, etc.) corresponding to a specific LSF. This will require multiple deterministic evaluations of the system as required by the sampling design scheme. A second-order polynomial will be generated by fitting the response data as discussed in Chapter 3. In all three sampling design schemes presented in Table 3.5, use Model 1 to generate the first IRS and the corresponding LSF so that the iterative process of FORM can be initiated. This indicates that the first IRS will be generated using SD without cross terms around the mean values of all the RVs. To execute the next step of FORM, as discussed in Section 2.10, an initial value of the reliability index β can be assumed, such as $\beta = 3$, to accelerate the convergence of the iterative process. Then the direction cosine value α_i of the ith RV at the checking point (mean value at the first iteration) will be computed. This step needs to be repeated until the estimates of α_i values converge for all RVs with a predetermined tolerance level. It may take a few iterations. Once the direction cosine values for all RVs converge, the LSF will be rewritten by keeping β as an unknown parameter. Solving the LSF equation, an updated value of β will be obtained. With the updated information on β and the applicable α_i values of all RVs at this stage, the coordinates of the new checking point or center point will be obtained. They are expected to be different, away from the mean values of all RVs. The second IRS will be generated around the new center point following the same procedures used to generate the first IRS. Then, following the same procedures, the second updated value of β will be available using the information on the second IRS and FORM, after the direction cosine values converge satisfying the predetermined convergence criterion. As before, the coordinates of another center point will be obtained, and the third IRS will be generated around the updated center point. The process of generating a new IRS using the information on the updated β value and the corresponding direction cosine values will continue until the β value converges satisfying the preselected convergence criterion. The final IRS will be generated around the latest center point calculated using the most recent information on β and the direction cosine values. To improve accuracy in estimating the probability of failure p_f, Model 2 or 3 can be used in the final iteration, as mentioned in Table 3.5. The β value obtained from the last IRS and the corresponding p_f value will give information on the underlying risk. This is essentially another updated version of REDSET.

The p_f value extracted by FORM using the explicit expression of an IRS and the corresponding LSF will complete the integration process of the finite element method (FEM), IRS, and FORM. It appears to be reasonable and straightforward. However, the basic REDSET concept, as discussed earlier, cannot be executed to estimate the risk of realistic large structural systems. To estimate p_f of CNDES accurately and efficiently using REDSET for routine applications, a series of intertwined difficult steps need to be executed at this stage as discussed next.

5.2.2 Increase Efficiency in Generating an IRS

An IRS represented by a second-order polynomial using various experimental design schemes just discussed is not expected to be efficient and accurate enough for the routine application of REDSET. It may require a very large number of nonlinear dynamic FE analyses (NDFEA) of the system to generate the response data for fitting an IRS, requiring continuous running of a computer for several years. The primary objective at this stage is to efficiently generate an explicit second-order expression for the IRS by conducting as few NDFEA as practicable.

5.2.3 Optimum Number of NDFEA Required for the Generation of an IRS

To improve efficiency in generating an IRS, reducing the total number of NDFEA required to implement the sampling design schemes presented in Section 5.2.1 will be of interest at this stage. As discussed in Chapter 3, the total number of NDFEA required for generating an IRS with k total number of RVs using sampling design schemes SD without cross terms, SD with cross terms, and CCD with cross terms will be $2k + 1$, $(k + 1)(k + 2)/2$, and $2^k + 2k + 1$, respectively. For a relatively small value of k, say $k = 10$, it will require 21, 66, and 1045 NDFEA, respectively. However, if $k = 100$, it will require 201, 5151, and $\approx 1.267 \times 10^{30}$ NDFEA to generate one IRS. As discussed earlier, it will require the generation of a few IRSs by updating the coordinates of the center point in the context of integrating the process with FORM. Thus, the total number of NDFEA will be referred to hereafter as TNDFEA to implement REDSET, which will be much more than the numbers shown earlier. As discussed in Chapter 3, the sampling design scheme CCD is expected to be more accurate compared to SD without and with cross terms but cannot be used for the reliability analysis of CNDES. All sampling design schemes discussed so far appear to be unattractive when k is relatively large. This is a major hurdle and must be addressed before moving forward. The obvious first strategy will be to reduce the value of k, the total number of RVs present in the formulation.

5.2.4 Reduction of Random Variables

Most real CNDES are expected to have a very large number of RVs, i.e. k will be large. However, it is well known that the propagations of uncertainties from the parameter to the system level are not equal for all the RVs. As discussed in Section 2.11, the uncertainty propagation characteristic of an RV from the element to the system level can be assessed in terms of its sensitivity index. The sensitivity of an RV X_i, α_{X_i} is the direction cosine of the unit normal variable at the checking or design point. The information on direction cosine values will be readily available from FORM analyses as discussed in detail in Section 5.2.1 or in Section 2.10. The efficiency of the algorithm can be improved significantly without sacrificing accuracy by considering some of the RVs with smaller sensitivity indexes to be deterministic at their mean values. Denoting the reduced number of RVs as k_R, the total number of RVs will be reduced from k to k_R. Based on the experience gained in dealing with SFEM (Huh and Haldar 2002, 2011; Farag et al. 2016), the computational efficiency in generating an IRS will be significantly improved with this reduction scheme requiring fewer numbers of TNDFEA without compromising the accuracy. Thus, k needs to be replaced by the reduced number of RVs k_R in all the sampling design schemes proposed in Chapter 3 or earlier in this chapter. The improvement in efficiency will be explained further with the help of examples later in this chapter.

5.3 Advanced Sampling Design Schemes

The reduction in the total number of RVs is a step in the right direction for improving efficiency, but a smaller value of k_R (for the sake of discussion, say less than 10) may not assure the accuracy required for a specific problem since the information on risk will be unknown at the initiation of the reliability study. To increase the robustness of REDSET, this issue needs further attention. All the sampling design schemes discussed so far were developed by combining two basic designs SD and CCD. Thus, the next important task will be to explore improving their efficiency, if possible, for wider applications without compromising accuracy. If successful, the added benefit will be the improvement of the robustness of REDSET. Several advanced factorial design (AFD) schemes are developed and their appropriateness in implementing REDSET is investigated.

5.4 Advanced Factorial Design Schemes

The generation of an IRS is iterative in nature. In the first attempt, several new AFD schemes are considered by combining the advantages of SD and CCD. Since the experimental design SD is the most efficient for generating a second-order polynomial without cross terms, it can be used for the first and the intermediate iterations to locate the failure region with reasonable accuracy. In the final iteration, either SD with cross terms or CCD with cross terms, Models 2 and 3, respectively, in Table 3.5 can be used to increase accuracy. Among several alternatives considered, the three most promising AFD schemes are discussed next.

Scheme 1 (S1): Use SD with second-order polynomial without cross terms throughout all the iterations; first, intermediates, and the final iteration. This scheme is expected to be the most efficient but least accurate in estimating the probability of failure.

Scheme 2 (S2): Use SD with second-order polynomial without the cross terms in the intermediate iterations and SD with a full second-order polynomial in the final iteration. Suppose n is the total number of IRSs necessary to extract reliability information using REDSET. To start this scheme, in the first iteration, an IRS will be generated at the mean values of all k RVs requiring $2k + 1$ NDFEA. From the second to $(n - 2)$ intermediate iterations, using the information of direction cosine values of all the RVs, k can be reduced to k_R for generating an IRS. This will require $(n-2)(2k_R+1)$ NDFEA. Using SD with a second-order polynomial with cross terms in the final iteration will require $(k_R+1)(k_R+2)/2$ NDFEA. Thus, the total NDFEA or TNDFEA required to extract reliability information using REDSET will be $(2k+1)+(n-2)(2k_R+1)+(k_R+1)(k_R+2)/2$.

Scheme 3 (S3): Use SD with second-order polynomial without the cross terms in the first and intermediate iterations, and CCD with a full second-order polynomial in the final iteration. In line with the discussions made for Scheme S2, the first IRS will be generated using $2k+1$ NDFEA. From the second to $(n-2)$ intermediate iterations, the total NDFEA required will be $(n-2)(2k_R+1)$. The last IRS generated using CCD will require $2^{k_R}+2k_R+1$ NDFEA. Thus, TNDFEA required using Scheme S3 to extract reliability information using REDSET will be $(2k+1)+(n-2)(2k_R+1)+2^{k_R}+2k_R+1$.

The implications of using these three AFD Schemes S1 to S3 are necessary at this stage. Two examples are considered to study the efficiency of these schemes, in terms of TNDFEA required to implement them in REDSET. The results are summarized in Table 5.1. A relatively small problem with $k = 10$ is considered in Example 1. In Example 2, a relatively large problem with $k = 100$ is considered. Suppose n indicates the number of IRS needed to extract reliability information using RED-SET. When $n = 3$, it indicates that the reliability index β converged after generating three IRSs (the initial, one intermediate, and the final); thus, the minimum value for n will be 3. TNDFEA values required for the two examples with for $n = 3$ and 5 are summarized in Table 5.1. When $n = 5$, it indicates that in addition to the initial and the final, three intermediate IRSs are required to be generated to implement REDSET.

TNDFEA reflects the efficiency of a scheme. The implications of different values for k, k_R, and n, on TNDFEA to implement REDSET are shown in Table 5.1. For Example 1 with $k = 10$, $k_R = 7$, and with two values of $n = 3$ and 5, the maximum value for TNDFEA using Scheme S3 will be 179 and 209, respectively. For the second example, for the same two values of n, the maximum value for TNDFEA will be 359 and 389, respectively. Essentially, it will take about two to three days of continuous running of a computer to extract reliability information, not several years as discussed earlier. For both examples, when Scheme S3 with CCD is used in the final iteration, it requires the maximum TNDFEA. As expected, the minimum required TNDFEA will be when Scheme S1 is used. Scheme S2 will require TNDFEA values between Schemes S1 and S3.

It will also be informative to compare the required number of sampling points shown in Table 3.4 and in Table 5.1. It will require an extremely large number of sampling points whenever CCD is used. For large values of k, the basic CCD design cannot be used. However, Scheme S3 is implementable even for large values of k if

Table 5.1 TNDFEA required to implement Schemes S1, S2, and S3.

(1)		(3)	(4)	Example 1 $k=10, k_R=7$ (5)		Example 2 $k=100, k_R=7$ (6)		TNDFEA Example 1 (7)		Example 2 (8)	
(2)				$n=3$	$n=5$	$n=3$	$n=5$	$n=3$	$n=5$	$n=3$	$n=5$
First iteration			$2k+1$	21	21	201	201				
Intermediate iteration			$(n-2)(2k_R+1)$	15	45	15	45				
Final iteration	AFD	S1	$2k_R+1$	15	15	15	15	51	81	231	261
		S2	$(k_R+1)(k_R+2)/2$	36	36	36	36	72	102	252	282
		S3	$2^{k_R}+2k_R+1$	143	143	143	143	179	209	359	389

it is replaced with k_R. As discussed before, Scheme S1 may not satisfy the accuracy requirement. Also, since Scheme S1 contains only the basic terms, its efficiency cannot be improved further. In search of the optimum sampling design scheme, it will not be an attractive candidate. AFD Schemes S2 and S3 require a few hundred of TNDFEA to implement REDSET. However, if k_R value is relatively large, both AFD Schemes S2 and S3 will be inefficient. They remained under consideration at this time, but attempts are made to improve their efficiency further.

5.5 Modified Advanced Factorial Design Schemes

To increase the efficiency of AFD Schemes S2 and S3 further, one reasonable option will be to explore the possibility of reducing the number of most sensitive variables in k_R further. After considering several options, the following modified AFD schemes appear to be very promising.

5.5.1 Modified Advanced Factorial Design Scheme 2 (MS2)

As discussed earlier, Scheme S2 consists of one center point, $2k_R$ axial points, and $k_R(k_R - 1)/2$ edge points. Referring to Eq. 3.9, it can be observed that the total number of cross terms required to generate a second-order polynomial using Scheme S2 will depend on the value of k_R. To improve the efficiency of Scheme S2 further without compromising the overall accuracy in the estimated risk, whether the cross terms for all k_R RVs are needed or not was investigated. To study this objective, k_R RVs were arranged in descending order of their sensitivity indexes. In the modified Scheme S2, denoted hereafter as Scheme MS2, the cross terms were added only for the RV with the highest sensitivity index. The cross terms for other RVs with smaller sensitivity indexes were added one by one in a sequence and the corresponding reliability index values were estimated until the reliability index β converged with a preassigned tolerance level. For k_R RVs, suppose the cross terms for m RVs are needed to be considered without compromising the convergence criterion of β and $m \leq k_R$. It can be shown that TNDFEA required to extract reliability information using Schemes S2 and MS2 will be $(2k + 1) + (n - 2)(2k_R + 1) + (k_R + 1)(k_R + 2)/2$ and $(2k + 1) + (2k_R + 1) + m(2k_R - m - 1)/2$, respectively.

The same two examples presented in Table 5.1 are considered again to document the improvement in the efficiency in terms of reducing the TNDFEA values when Scheme MS2 is used. For Example 1, with $k = 10$, $k_R = 7$, $n = 5$, and $m = 3$, TNDFEA will be 90 using Scheme MS2 instead of 102 using Scheme S2. For Example 2, with $k = 100$, $k_R = 7$, $n = 5$, and $m = 3$, the corresponding TNDFEA using Schemes MS2 and S2 will be 276 and 282, respectively. The information is summarized in Table 5.2.

Table 5.2 Advanced factorial designs using Schemes S1, MS2, and MS3.

(1)	(2)	(3)	(4)	Example 1 (5) $k=10$, $k_R=7$, $m=3$		Example 2 (6) $k=100$, $k_R=7$, $m=3$		TNDFEA (7) Example 1		TNDFEA (8) Example 2	
				$n=3$	$n=5$	$n=3$	$n=5$	$n=3$	$n=5$	$n=3$	$n=5$
First iteration			$2k+1$	21	21	201	201				
Intermediate iteration			$(n-2)(2k_R+1)$	15	45	15	45				
Final iteration	AFD	S1	$2k_R+1$	15	15	15	15	51	81	231	261
		MS2	$2k_R+1+m(2k_R-m-1)/2$	30	30	30	30	66	90	246	276
		MS3	2^m+2k_R+1	23	23	23	23	59	89	239	269

For both examples, the modification indicates improvements in efficiency without compromising accuracy. It needs to be pointed out that TNDFEA required to implement REDSET using Scheme MS2 will be smaller than Scheme S2 when $m < k_R$.

5.5.2 Modified Advanced Factorial Design Scheme 3

The final iteration of Scheme S3 using CCD consists of one center point, $2k_R$ axial points, and 2^{k_R} factorial points. Similar to Scheme MS2, the cross terms and the necessary sampling points are considered only for the most significant RVs in k_R without compromising the accuracy in estimating the reliability index β. However, to generate the IRS of interest, a regression analysis is required. The reduction of factorial points less than the number of coefficients will cause ill-conditioning of the regression analysis due to the lack of data. To avoid ill-conditioning, the number of unknown coefficients always needs to be less than the number of sampling points. To satisfy this requirement, just the cross terms for m most significant variables without causing ill-conditioning are considered. It will be denoted hereafter as modified Scheme S3 or Scheme MS3. It can be shown that TNDFEA required to extract reliability information using Schemes S3 and MS3 will be $(2k + 1) + (n - 2)(2k_R + 1) + 2^{k_R} + 2k_R + 1$ and $(2k + 1) + (n - 2)(2k_R + 1) + 2^m + 2k_R + 1$, respectively. The same two examples discussed earlier in Table 5.1 are considered again to document the improvement in efficiency when Scheme MS3 is used. For Example 1, with $k = 10$, $k_R = 7$, $n = 5$, and $m = 3$, TNDFEA required will be 89 and 209 when Schemes MS3 and S3, respectively are used. For Example 2, with $k = 100$, $k_R = 7$, $n = 5$, and $m = 3$, the corresponding TNDFEA will be 269 and 389, respectively. The information is also summarized in Table 5.2. For both examples, the modifications indicate improvement. Again, as stated earlier for Scheme S2, TNDFEA required to implement REDSET using Scheme MS3 will be smaller than Scheme S3 when $m < k_R$. The most encouraging observation in dealing with Scheme S3 or MS3 is that the required TNDFEA will be few hundreds and not of the order of 10^{30}, indicating REDSET is implementable to estimate the reliability of realistic CNDES with very large number of RVs.

Schemes S1, S2, S3, MS2, and MS3 will be collectively grouped under AFD in the subsequent discussions. As discussed earlier, the major objective is to extract reliability information of CNDES by conducting as few TNDFEA as possible without compromising the accuracy. Discussions made so far indicates that the required total number of TNDFEA will depend on the values of k, k_R, n, and m.

5.6 Optimum Number of TNDFEA Required to Implement REDSET

The discussions made in the previous sections open up another important topic of reliability estimation using REDSET with the optimum number of TNDFEA. However, before making any conclusion on this important question, another factor needs to be considered. The information summarized in Table 5.2 indicates that Scheme MS3 requires the minimum TNDFEA, in some cases. It generates the final IRS using CCD with second-order polynomial with cross terms with the help of the regression analysis of the response data. The regression analysis may not incorporate the information on uncertainty embedded in the data with sufficient efficiency and there is room for improvement. Finding an alternative to the regression analysis may provide such improvement. Attempts of developing several new AFDs for generating an IRS without using the regression analysis may address the issue related to optimum TNDFEA. This discussion will address the third deficiency as discussed in Section 3.5.

5.7 Improve Accuracy of Scheme MS3 Further – Alternative to the Regression Analysis

It is now necessary to investigate how to improve the accuracy of Scheme MS3. Two improvement strategies are considered as discussed next.

5.7.1 Moving Least Squares Method

In implementing Scheme MS3, the experimental design scheme CCD is used to generate the final IRS using the least squares method (LSM)-based regression analysis. An IRS generated in this way fits the data in an average sense. Any function generated using the regression analysis is expected to have a considerable amount of uncertainty about the average trend as briefly discussed in Section 3.7. It is essentially a global approximation technique in which all the sample points are assigned the same weight factor of one. This assumption may not be the most appropriate for the class of problems under consideration. The weight factors should decay as the distances between the sampling points and the IRS increase. The moving least squares method (MLSM) (Azizsoltani and Haldar 2017) can be an alternative to promote this hypothesis. In MLSM, an IRS is generated using the generalized least squares technique with different weights factors for the different sampling points (Kang et al. 2010; Li et al. 2012; Taflanidis and Cheung 2012; Chakraborty and Sen 2014). This will ensure that the sampling points located closer to the IRS will have higher weight factors instead of equal used in the conventional LSM-based algorithm.

5.7.2 Concept of Moving Least Squares Method

Even at the risk of repetition, it is important to state that the generation of an IRS using MLSM only relates to Schemes S3 and MS3 where sampling design scheme CCD is used to generate the final IRS. Instead of using the regression analysis with an equal weight factor using LSM, MLSM can be used with decaying weight factors as the distances between the sampling points and the IRS increase. To implement MLSM, different weight factors for the sampling points need to be estimated. Different types of weight factors are reported in the literature considering the form of a polynomial of interest, including constant, linear, quadratic, fourth-order polynomial, exponential, etc. In this study, the following fourth-order polynomial is used to estimate the weight factor:

$$w(d_i) = 1 - 6\,r^2 + 8\,r^3 - 3\,r^4; \text{if } r \le 1.0$$

$$= 0; \text{if } r > 1.0 \tag{5.1}$$

where $r = d_i/R_i$ is the ratio of the distance and region of influence, d_i is the distance between the ith sampling point and the point on the IRS where the weight factor needs to be evaluated, and R_i is the region of influence of that sampling point. In order to avoid ill-conditioning of the matrix, R_i is selected in such a way that it includes sufficient sampling points. It is chosen as twice the distance between the center point and the furthest sampling point (Bhattacharjya and Chakraborty 2011).

Assuming that there are a total of N sampling points, the regression model can be expressed in the matrix form as:

$$\mathbf{Y} = \mathbf{Bb} + \varepsilon \tag{5.2}$$

where $\mathbf{Y}$ is a vector of size $N \times 1$ containing N observations for generating an IRS; $\mathbf{b}$ is the unknown coefficient vector to be calculated using MLSM, $\mathbf{B}$ is a matrix of size $N \times p$ containing the coefficients of the second-order polynomial for an IRS, p is the number of unknown coefficients to be estimated for an IRS, and ε is a vector of size $N \times 1$ containing the errors or residuals. The $\mathbf{B}$ matrix can be shown to be:

$$B = \begin{bmatrix} 1 & x_{11} & x_{21} & \cdots & x_{k1} & x_{11}^2 & x_{21}^2 & \cdots & x_{k1}^2 & x_{11}x_{21} & x_{11}x_{31} & \cdots & x_{11}x_{k1} & \cdots & x_{k-1,1}x_{k1} \\ 1 & x_{12} & x_{22} & \cdots & x_{k2} & x_{12}^2 & x_{22}^2 & \cdots & x_{k2}^2 & x_{12}x_{22} & x_{12}x_{32} & \cdots & x_{12}x_{k2} & \cdots & x_{k-1,2}x_{k2} \\ \cdot & \cdot & \cdot & & \cdot & \cdot & \cdot & & \cdot & \cdot & \cdot & & \cdot & & \cdot \\ \cdot & \cdot & \cdot & & \cdot & \cdot & \cdot & & \cdot & \cdot & \cdot & & \cdot & & \cdot \\ \cdot & \cdot & \cdot & & \cdot & \cdot & \cdot & & \cdot & \cdot & \cdot & & \cdot & & \cdot \\ 1 & x_{1N} & x_{2N} & \cdots & x_{kN} & x_{1N}^2 & x_{2N}^2 & \cdots & x_{kN}^2 & x_{1N}x_{2N} & x_{1N}x_{3N} & \cdots & x_{1N}x_{kN} & \cdots & x_{k-1,N}x_{kN} \end{bmatrix}$$

$$\tag{5.3}$$

where x_{ij} is the observation of the ith RV in the jth data set. For k RVs, p can be estimated as $(k + 1)(k + 2)/2$, as discussed earlier in Chapter 3. In order to estimate the unknown coefficient vector $\mathbf{b}$ at the checking point, the weighted sum of the squared residuals at all the sampling points needs to be minimized. The weighted sum S can be expressed as:

$$S = (\mathbf{Y} - \mathbf{Bb})^{\mathrm{T}} \mathbf{W} (\mathbf{Y} - \mathbf{Bb}) \tag{5.4}$$

where $\mathbf{W}$ is a diagonal matrix containing the weight factors and can be expressed as:

$$\mathbf{W} = \begin{bmatrix} w(d_1) & 0 & \cdots & 0 \\ 0 & w(d_2) & \cdots & 0 \\ \cdot & \cdot & \cdot & \cdot \\ \cdot & \cdot & \cdot & \cdot \\ \cdot & \cdot & \cdot & \cdot \\ 0 & 0 & \cdots & w(d_n) \end{bmatrix} \tag{5.5}$$

where $w(d_i)$ is the weight factor corresponding to ith sampling point. Minimizing the weighted sum S in Eq. 5.4, the unknown coefficient vector $\mathbf{b}$ can be expressed as (Haldar and Mahadevan 2000a):

$$\mathbf{b} = \left[\mathbf{B}^{\mathrm{T}} \mathbf{W} \mathbf{B} \right]^{-1} \mathbf{B}^{\mathrm{T}} \mathbf{W} \mathbf{Y} \tag{5.6}$$

Using the information on the improved regression coefficients using the MLSM concept, the final IRS can be generated. Using the modified IRS and the corresponding LSF, and the reliability index β can be calculated using FORM. The extracted reliability information is expected to be different than when the ordinary regression analysis with LSM is used. For ease of discussion, this design will be denoted hereafter as Scheme S3-MLSM.

Scheme S3-MLSM is compared with other schemes presented so far in Table 5.3 in terms of the required TNDFEA to implement REDSET without considering accuracy. The results indicate that Schemes S3-MLSM and MS3 will require identical TNDFEA for both examples to generate the required IRS. Since IRS using MLSM is expected to be more appropriate, the reliability index estimated this way is expected to be more accurate. However, the fact remains that MLSM estimates the coefficients of the polynomial in the IRS in an average sense, an alternative to the regression analysis was explored further as discussed next.

5.7.3 Improve Efficiency Further to the Moving Least Squares Method

In line with the discussions made in proposing Scheme MS3, an attempt was made to improve the efficiency of Scheme S3-MLSM further. It will be denoted hereafter as Scheme MS3-MLSM. Using this scheme, an IRS is generated using m numbers of the most significant RVs instead of the reduced number of RVs k_R as discussed in Section 5.5. TNDFEA required for Scheme MS3-MLSM will be lower than Scheme

Table 5.3 Required TNDFEA to implement REDSET using Schemes S1, S2, MS2, S3, S3-MLSM, and MS3-MLSM.

(1)	(2)	(3)	(4)	Example 1	Example 2	TNDFEA	
				$k=10, k_R=7, n=3, m=2$ (5)	$k=100, k_R=7, n=3, m=2$ (6)	Example 1 (7)	Example 2 (8)
First iteration			$2k+1$	21	201	51	231
Intermediate iteration			$(n-2)(2k_R+1)$	15	15		
Final iteration	AFD	S1	$2k_R+1$	15	15	51	231
		S2	$(k_R+1)(k_R+2)/2$	36	36	72	252
		MS2	$2k_R+1+m(2k_R-m-1)/2$	26	26	62	242
		S3	$2^{k_R}+2k_R+1$	143	143	179	359
		MS3	2^m+2k_R+1	19	19	55	235
	MLSM	S3-MLSM	$2^{k_R}+2k_R+1$	143	143	179	359
		MS3-MLSM	2^m+2k_R+1	19	19	55	235

S3-MLSM for both examples as shown in Table 5.3, since m is expected to be smaller than k_R, in most cases.

As discussed earlier, the basic deficiency in Schemes S3, MS3, S3-MLSM, and MS3-MLSM is that an IRS is generated using an average sense; it will not pass through all the response data generated. To address this deficiency, a Kriging method (KM)-based alternative (Krige 1951) is explored, as discussed next.

5.8 Generation of an IRS Using Kriging Method

KM is a surrogate metamodel capable of generating an IRS where all sample points pass through the polynomial to be generated. For given sampling data, differences between the curves generated using the regression analysis and KM are shown in Figure 5.1. It indicates that an IRS generated using KM will pass through all the sample points. It will also predict the response information between two sample points more accurately where responses were not available.

KM was originally developed to improve the accuracy of tracing gold in ores. It is a geostatistical method of interpolation with the prior assumption that the function to be estimated is a Gaussian process. In the context of generating an IRS, the Gaussian assumption is acceptable. KM depends on the idea that the near sample points should get more weight (Lichtenstern 2013). The KM concept uses a weighted average of the responses obtained by deterministic analyses to generate an IRS and its derivatives at unobserved points. Two basic desirable characteristics of KM are that an IRS generated by it is uniformly unbiased and its prediction error

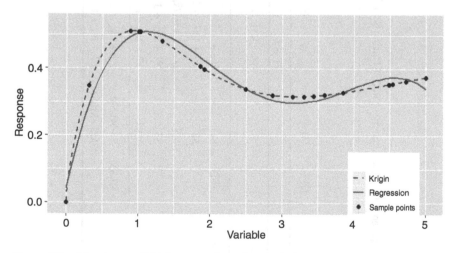

Figure 5.1 Advantages of Kriging over regression analysis.

is less than all other possible forms. These features make it the best linear unbiased surrogate for an IRS (Wackernagel 2013).

Because of its many desirable characteristics, several types of KM are reported in the literature to address available problem-specific information. They include simple (when the known mean value of the simulated function can be estimated from the underlying physics of the problem), ordinary (when the constant mean value of the simulated function is not available before starting the process), and universal Kriging (UK) (when a drift function needs to be considered over the spatial region) (Azizsoltani and Haldar 2017). They are briefly discussed below.

5.8.1 Simple Kriging

Simple Kriging assumes the mean value of random function $Z(\mathbf{X})$ is known and constant. This is the reason simple Kriging is also known as Kriging with a known mean. Furthermore, $Z(\mathbf{X})$ is supposed to be second-order stationary with a constant mean value, i.e. $E[Z(\mathbf{X})] = \mu(\mathbf{X}) = \mu$. The covariance function, also called covariogram, $C(\mathbf{h})$, depends only on the separating vector, $\mathbf{h}$, i.e. $C(\mathbf{h}) \equiv \text{Cov}[Z(\mathbf{X}), Z(\mathbf{X} + \mathbf{h})]$. A process is said to be second-order stationary (or weakly stationary) if (i) the mean of the process does not depend on $\mathbf{X}$: $E[Z(\mathbf{X}] = \mu$, (ii) the variance of the process does not depend on $\mathbf{X}$: $E[(Z(\mathbf{X}) - \mu)^2] = \mu$, and (iii) the covariance $C(\mathbf{h})$ only depends on the separation between two points of the process and does not depend on $\mathbf{X}$: $C(\mathbf{h}) \equiv \text{Cov}[Z(\mathbf{X}), Z(\mathbf{X} + \mathbf{h})] = E[Z(\mathbf{X})Z(\mathbf{X} + \mathbf{h})] - \mu^2$. An IRS represented by $\hat{g}(\mathbf{X})$ can be used as the simple Kriging predictor at $\mathbf{X_0}$:

$$\hat{g}(\mathbf{X_0}) \equiv \mu + \sum_{i=1}^{r} \omega_i[Z(\mathbf{X}_i) - \mu] = \mu + \boldsymbol{\omega}^{\mathrm{T}}(\mathbf{Z} - \mu\mathbf{1}) \tag{5.7}$$

where $\omega_i \in R$, $i = 1, 2, ..., r$ denotes unknown weights corresponding to $Z(\mathbf{X}_i) - \mu$ and $\boldsymbol{\omega} \equiv (\omega_1, ..., \omega_n)^{\mathrm{T}}$ is the weight vector. By applying the two conditions of minimizing the error and unbiasedness, it can be shown that:

$$\sum_{j=1}^{r} \omega_j C(\mathbf{X}_i - \mathbf{X}_j) = C(\mathbf{X}_i - \mathbf{X_0}); i = 1,2,...,r \tag{5.8}$$

Thus, all the weights can be calculated as:

$$\begin{bmatrix} \omega_1 \\ \vdots \\ \omega_r \end{bmatrix} = \begin{bmatrix} C(\mathbf{X}_1 - \mathbf{X}_1) & \cdots & C(\mathbf{X}_1 - \mathbf{X}_r) \\ \vdots & C(\mathbf{X}_i - \mathbf{X}_i) & \vdots \\ C(\mathbf{X}_r - \mathbf{X}_1) & \cdots & C(\mathbf{X}_r - \mathbf{X}_r) \end{bmatrix}^{-1} \begin{bmatrix} C(\mathbf{X}_1 - \mathbf{X_0}) \\ \vdots \\ C(\mathbf{X}_r - \mathbf{X_0}) \end{bmatrix} \tag{5.9}$$

or

$$\boldsymbol{\omega} = \boldsymbol{\Sigma}^{-1}\mathbf{C_0}$$

where $\mathbf{C_0} \equiv [C(\mathbf{X}_1 - \mathbf{X_0}), ..., C(\mathbf{X}_r - \mathbf{X_0})]^{\mathrm{T}} \in R^r$ denotes the covariance between samples and location of interest. Also, $\boldsymbol{\Sigma} \in R^{r \times r}$ is the covariance matrix of $\mathbf{Z}$. Using the information on $\boldsymbol{\omega}$, the simple Kriging predictor $\hat{g}(\mathbf{X_0})$ can be calculated.

5.8.2 Ordinary Kriging

Ordinary Kriging considers the global mean constant; however, unlike simple Kriging, does not consider mean as known, which makes ordinary Kriging more realistic than simple Kriging. This method also assumes $Z(\mathbf{X})$ as an intrinsically stationary random function based on a predefined variogram function $\gamma(\mathbf{h})$, i.e.

$$\gamma(\mathbf{h}) \equiv \frac{1}{2} \text{Var} \left[Z(\mathbf{X} + \mathbf{h}) - Z(\mathbf{x}) \right] = \frac{1}{2} E\left[(Z(\mathbf{X} + \mathbf{h}) - Z(\mathbf{X}))^2 \right] \tag{5.10}$$

The variogram function will be discussed in more detail in Section 5.8.4. An intrinsically stationary process should satisfy the following conditions: (i) the mean of the process is constant: $E[Z(\mathbf{x}) - Z(\mathbf{x} + \mathbf{h})] = 0$, (ii) the variance of the process does not depend on $\mathbf{X} : E[(Z(\mathbf{X}) - \mu)^2] = \mu$, and (iii) the variance of the process is finite and only depends on the separation between two points of the process: $\text{Var}[Z(\mathbf{X}) - Z(\mathbf{X} + \mathbf{h})] < \infty$. The ordinary Kriging predictor at the required location $\mathbf{X_0}$ can be formulated as:

$$\hat{g}(\mathbf{X_0}) \equiv \sum_{i=1}^{r} \omega_i Z(\mathbf{X}_i) = \omega^{\text{T}} \mathbf{Z} \tag{5.11}$$

In order to minimize the expected error, the sum of the weights should equal one:

$$\sum_{i=1}^{r} \omega_i = \mathbf{1} \text{ or } \omega^{\text{T}} \mathbf{1} = 1 \tag{5.12}$$

Minimizing the error considering the aforementioned constraint and using the Lagrange multiplier $\lambda \equiv (\lambda_1, ..., \lambda_r)^{\text{T}} \in R^r$ results in:

$$\Gamma \omega + \lambda \mathbf{1} = \gamma_0 \tag{5.13}$$

where $\Gamma \in R^{r \times r}$ is symmetric variogram matrix, $\Gamma_{\gamma i, j} \equiv \gamma_\gamma(\mathbf{X}_i - \mathbf{X}_j)$, $i, j = 1, 2, ..., r$, and $\gamma_0 \equiv [\gamma(\mathbf{X}_1 - \mathbf{X}_0), ..., \gamma(\mathbf{X}_r - \mathbf{X}_0)]^{\text{T}} \in R^r$. Assuming Γ is invertible, the weights can be calculated as:

$$\omega = \Gamma^{-1} \left[\gamma_0 - \mathbf{1} \left(\frac{\mathbf{1}\Gamma^{-1}\gamma_0 - 1}{\mathbf{1}^T \Gamma^{-1} \mathbf{1}} \right) \right] \tag{5.14}$$

With the availability of ω, $\hat{g}(\mathbf{X_0})$ at the location of $\mathbf{X_0}$ can be calculated.

5.8.3 Universal Kriging

An IRS at unobserved points predicated by UK is a linear weighted sum of the responses calculated by the FE analyses in the context of this study and can be expressed as (Webster and Oliver 2007):

$$\hat{g}(\mathbf{X}) = \sum_{i=1}^{r} \omega_i Z(\mathbf{X}_i) \tag{5.15}$$

where $\omega_i \in R$, $i = 1, 2, ..., r$ are unknown weights to be determined corresponding to the vector of the observed points $\mathbf{Z} \equiv [Z(\mathbf{X}_i), i = 1, 2, ..., r]$ and r is the number of the observed points or the number of FE analyses. The vector of the observed points $\mathbf{Z}$ consists of the responses at experimental sampling points.

In UK, the Gaussian process $Z(\mathbf{X})$ is the summation of a deterministic function $u(\mathbf{X})$ and a residual random function $Y(\mathbf{X})$. It can be mathematically expressed as:

$$Z(\mathbf{X}) = u(\mathbf{X}) + Y(\mathbf{X}) \tag{5.16}$$

where $u(\mathbf{X})$ is a second-order polynomial with cross terms, as in Eq. 3.9. The residual random function is intrinsically stationary with zero mean. Its spatial continuity is described by the underlying variogram function γ_Y, a very important parameter in generating a KM-based IRS and will be discussed in more detail in Section 5.8.4.

The variogram function is generated based on the assumption that the difference between responses at two sample points only depends on their relative locations. To generate the appropriate variogram function, three main steps are (i) the generation of the variogram cloud, (ii) the experimental variogram, and (iii) an appropriate variogram model to fit the experimental variogram. The graphical representation of the dissimilarity function against distance is called the variogram cloud. Dissimilarity function, $\gamma^*(l_i)$, can be expressed in terms of responses at two sample points separated by a distance l_i as:

$$\gamma^*(l_i) = \frac{1}{2}[Z(x_i + l_i) - Z(x_i)]^2 \tag{5.17}$$

where x_i is the coordinate of the experimental sample point in the axis of ith RV. It is symmetric with respect to l_i and can be plotted against the absolute value of l_i for each RV axis. When CCD is used in the final step of REDSET at the nth iteration, both nth and $(n-1)$th iterations are expected to be in the failure region. Since a large data set is preferable to increase accuracy, responses at both iterations are used for generating the variogram cloud. This is expected to produce large sets of dissimilarity values for a particular distance l_i.

An experimental variogram is the average of the dissimilarities with the same distance l_i. Several valid parametric variogram models are considered for fitting to the experimental variogram, including Nugget effect model, Bounded linear model, Spherical model, and Exponential model. After fitting the valid parametric models using the least square and weighted least square regression techniques, it was observed that the family of stable anisotropic variogram models with weighted least square regression indicates the best fit with the coefficient of determination close to 1. They can be represented as (Wackernagel 2013):

$$\gamma_Y(\mathbf{l}) = b\left\{1 - \exp\left[\sum_{i=1}^{k_R} -\frac{|l_i|^q}{a_i}\right]\right\} \text{ for } 0 < q < 2 \tag{5.18}$$

where $\mathbf{1}$ is a vectorial form of the l_i components a_i and b are the unknown coefficients to be estimated, and k_R is the reduced number of sensitive RVs. More discussions on these coefficients and on variograms are made in Section 5.8.4.

Considering the universality conditions on the unknown weights imposes uniform unbiasedness on the UK predictor as (Cressie 2015):

$$\mathbf{F}^T \boldsymbol{\omega} = \mathbf{f}_0 \tag{5.19}$$

where $\boldsymbol{\omega} \equiv \omega_i$ is the unknown weight vector, $\mathbf{F} \equiv f_p(\mathbf{X}_i)$ is the matrix of ordinary regression, $\mathbf{f}_0 \equiv f_p(\mathbf{X}_0)$ is the vector of ordinary regression, $\mathbf{X}_i$ is the ith sampling point, $\mathbf{X}_0$ is the coordinate in which the response of the system should be predicted, and $f_p(\mathbf{X}_i)$ is the ordinary regression function for a second-order polynomial with cross terms for this study. Following the ordinary regression method, for each regressor variable $\mathbf{X}_i$, r sets of data consisting of P observations are gathered (Haldar and Mahadevan 2000a). Therefore, $\mathbf{F}$ is a matrix of size $r \times (P+1)$, $\mathbf{f}_0$ is a vector of size $(P+1) \times 1$, and $\boldsymbol{\omega}$ is a vector of size $r \times 1$.

Many sets of weights can satisfy the universality conditions but just one set has the minimum prediction error variance. An optimization process using Lagrange multipliers can be used to find the unknown weights. The optimality condition can be expressed as:

$$\boldsymbol{\Gamma}_Y \boldsymbol{\omega} + \mathbf{F} \boldsymbol{\lambda} = \boldsymbol{\gamma}_{Y,0} \tag{5.20}$$

where $\boldsymbol{\Gamma}_Y \equiv \gamma_Y(\mathbf{X}_i - \mathbf{X}_j)$, $i, j = 1, ..., r$ is the matrix of symmetric residual variogram, $\boldsymbol{\gamma}_{Y,0} \equiv \gamma_Y(\mathbf{X}_i - \mathbf{X}_0)$, $i = 1, ..., r$ is the vector of residual variogram vector, and $\boldsymbol{\lambda} \equiv (\lambda_0, \lambda_1, ..., \lambda_P)^T$ is the vector of unknown Lagrange multiplier to be determined. Therefore, $\boldsymbol{\Gamma}_Y$ is a matrix of $r \times r$, $\boldsymbol{\gamma}_{Y,0}$ is a vector of $r \times 1$, and $\boldsymbol{\lambda}$ is a vector of $(P+1) \times 1$.

The system of equations satisfying the universality and optimality conditions can be solved simultaneously. The system has a unique solution if all columns of matrix $\mathbf{F}$ are linearly independent (Wackernagel 2013). Required weights and Lagrange multipliers can be expressed as:

$$\begin{pmatrix} \boldsymbol{\omega} \\ \boldsymbol{\lambda} \end{pmatrix} = \begin{pmatrix} \boldsymbol{\Gamma}_Y & \mathbf{F} \\ \mathbf{F}^T & \mathbf{0} \end{pmatrix}^{-1} \begin{pmatrix} \boldsymbol{\gamma}_{Y,0} \\ \mathbf{f}_0 \end{pmatrix} \tag{5.21}$$

This system of equations can be solved using an iterative method. However, the only required part of the solution for generating an IRS is the unknown weights. It can be derived in closed form as:

$$\boldsymbol{\omega} = \boldsymbol{\Gamma}_Y^{-1} \left[\boldsymbol{\gamma}_{Y,0} - \mathbf{F} \left(\mathbf{F}^T \boldsymbol{\Gamma}_Y^{-1} \mathbf{F} \right)^{-1} \left(\mathbf{F}^T \boldsymbol{\Gamma}_Y^{-1} \boldsymbol{\gamma}_{Y,0} - \mathbf{f}_0 \right) \right] \tag{5.22}$$

The response surface $\hat{g}(\mathbf{X})$ in Eq. 5.15 can estimate the response of the system in each unsampled point. Thus, UK can generate the IRS as (Lichtenstern 2013):

$$\hat{g}(\mathbf{X}) = \left[\boldsymbol{\gamma}_{Y,0} - \mathbf{F}\left(\mathbf{F}^T\boldsymbol{\Gamma}_Y^{-1}\mathbf{F}\right)^{-1}\left(\mathbf{F}^T\boldsymbol{\Gamma}_Y^{-1}\boldsymbol{\gamma}_{Y,0} - \mathbf{f}_0\right)\right]^T\boldsymbol{\Gamma}_Y^{-1}\mathbf{Z} \qquad (5.23)$$

An IRS obtained by Eq. 5.23 will give an explicit expression for an IRS using UK and the corresponding explicit expression for the required LSF will be available. Then, FORM can be used to extract reliability information. All the required steps are discussed in detail in (Sen et al. 2016; Azizsoltani and Haldar 2017).

For CDNES considered in this study, the mean values of spatial data are expected to depend on the location of the sampling points. Considering three alternatives of Kriging; simple, ordinary, and universal, since UK is capable of incorporating external drift functions as supplementary variables (Hengl 2009), it is used in this study. This will produce an IRS that will be the best linear unbiased predictor in the presence of drift in the mean.

5.8.4 Variogram Function

In implementing UK discussed earlier, the most important task is generating information on the required variogram function. Unfortunately, the underlying variogram function is unknown and should be estimated. As discussed earlier, to generate it, the information on the variogram cloud, the experimental variogram, and an appropriate variogram model to fit the experimental variogram must be available. The graphical representation of the dissimilarity function as a function of distance is called the variogram cloud. The dissimilarity function is defined by Eq. 5.17 where the responses at two sample points separated by a distance l_i in direction of the ith RV. The next step is creating the variogram cloud based on the experimental sampling points which is a plot of dissimilarity as a function of distance. It provides a vague picture of the relationship between the points in space. Since more than one observation for a specific distance is possible, an experiential or empirical variogram needs to be defined. It is a function of the average dissimilarity between any two observations as a function of the separating distance. Finally, a parametric variogram model needs to be used to fit the experimental variogram. Many models are reported in the literature, including Nugget-effect, bounded linear, spherical, exponential, Gaussian, etc. Many methods including the least squares, maximum likelihood, minimum norm quadratic, etc. can be used to fit the data to these models. Obviously, the accuracy of the experimental variogram increases when more data are used to generate it. It should be mentioned the covariance function (or covariogram) $C(\mathbf{h})$ and variogram $\gamma(\mathbf{h})$ are equivalent for a bounded variogram, i.e.

$$C(\mathbf{h}) = C(\mathbf{0}) - \gamma(\mathbf{h}) = \gamma(\infty) - \gamma(\mathbf{h}) \qquad (5.24)$$

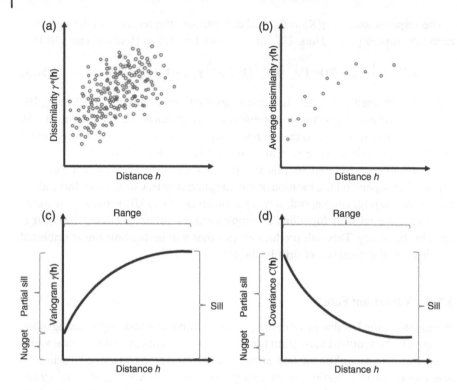

Figure 5.2 The schematic view of (a) variogram cloud; (b) experiential variogram; (c) variogram; and (d) covariogram.

Figure 5.2 schematically summarizes the three steps required to generate a variogram. In Figure 5.2a, b, the variogram cloud and experimental variogram, respectively, are shown. Variogram and covariogram are shown in Figure 5.2c, d, respectively, to identify several features. Nugget is the difference between $\gamma(\mathbf{0})$ and $\lim_{|\mathbf{h}| \to 0} \gamma(\mathbf{h})$, which is used to model discontinuity at the origin of the variogram. The value of $\lim_{|\mathbf{h}| \to \infty} \gamma(\mathbf{h})$ is defined as Sill. The partial sill is the difference between sill and nugget. Range is the distance when variogram reaches its maximum value. In the case of asymptotic functions, range is the distance when variogram reaches its maximum value or 95% of the maximum value for asymptotic functions.

The variogram function represented by Eq. 5.18 is used. Using the discussions just made, the three parameters a_i, b, and k_R in the equation can now be defined. All the models discussed earlier asymptotically approach the b value, commonly known as sill parameter. Parameter a_i, called the range parameter, represents the ith orthogonal direction range at which the variogram $\gamma_Y(\mathbf{1})$ almost passes through the 95% of the sill value. k_R is the reduced number of sensitive RVs, as discussed earlier. Since Eq. 5.18 with $q = 2$ is infinitely differentiable at the origin which makes it unrealistic, it is not considered in this study. It is observed in this study that the results are not very sensitive around $q = 2$. Finally, variograms are generated for each application using all the models discussed earlier and the model with the highest coefficient of determination (Haldar and Mahadevan 2000a) is selected to generate an IRS using Eq. 5.15.

It is to be noted that since KM uses the same number of total sample points TNDFEA to generate an IRS using CCD with Sampling Schemes S3, MS3, S3-MLSM, and MS3-MLSM, REDSET cannot be implemented if k_R is not relatively small. There is room for further improvement to increase efficiency, as discussed next.

5.8.5 Scheme S3 with Universal Kriging Method

Based on the discussions made in the previous sections, considering efficiency, accuracy, and robustness, another experimental sampling scheme can be developed by using the basic Scheme S3 with an IRS generated using KM instead of LSM-based regression analysis. This new sampling scheme will be denoted hereafter as Scheme S3-KM.

5.8.6 Scheme MS3 with Modified Universal Kriging Method

If the number of RVs is large even after selecting the reduced number of sensitive RVs k_R, the use of CCD in Scheme S3 or S3-KM in the final iteration will require a large number of TNDFEA reducing its efficiency. Similar to Scheme MS3, another experimental sampling scheme can be developed by using the basic Scheme MS3 with an IRS generated using KM instead of LSM-based regression analysis. This new sampling scheme will be denoted hereafter as Scheme MS3-KM. For Scheme MS3-KM, the total number of required FE analyses will remain the same as that of Scheme MS3.

5.9 Comparisons of All Proposed Schemes

The efficiency of all nine schemes discussed in the previous sections, namely Schemes S1, S2, MS2, S3, MS3, S3-MLSM, MS3-MLSM, S3-KM, and MS3-KM, can be studied with the help of two examples. The same two examples discussed earlier with $k = 10$ and $k = 100$ are considered. After generating the first IRS with k RVs, the reduced number of sensitive RVs k_R is considered to be 7 in both examples. TNDFEA values given in Table 5.4 are estimated based on the assumption that the reliability index converged after generating three IRSs or $n = 3$, and $m = 2$ indicating the cross terms for the modified sampling schemes. The total number of NDFEA (TNDFEA) required to implement Schemes S1, S2, MS2, S3-MLSM, MS3-MLSM, S3, MS3, KM, and MKM are shown in Table 5.4 Columns 7 and 8 for Examples 1 and 2, respectively.

Table 5.4 Required TNDFEA to implement REDSET with all nine schemes.

(1)	(2)	(3)	(4)	Example 1	Example 2	TNDFEA	
				$k=10, k_R=7, n=3, m=2$ (5)	$k=100, k_R=7, n=3, m=2$ (6)	Example 1 (7)	Example 2 (8)
First iteration			$2k+1$	21	201	51	231
Intermediate iteration			$(n-2)(2k_R+1)$	15	15	15	15
Final iteration	AFD	S1	$2k_R+1$	15	15	51	231
		S2	$(k_R+1)(k_R+2)/2$	36	36	72	252
		MS2	$2k_R+1+m(2k_R-m-1)/2$	26	26	62	242
		S3	$2^{k_R}+2k_R+1$	143	143	179	359
		MS3	2^m+2k_R+1	19	19	55	235
	MLSM	S3-MLSM	$2^{k_R}+2k_R+1$	143	143	179	359
		MS3-MLSM	2^m+2k_R+1	19	19	55	235
	Kriging	S3-KM	$2^{k_R}+2k_R+1$	143	143	179	359
		MS3-KM	2^m+2k_R+1	19	19	55	235

Several important observations can be made from the comparisons shown in Table 5.3. It should be noted that in the first and all intermediate iterations of generating IRSs, experimental design SD without cross terms is used. For the small problem with $k = 10$, the total NDFEA required to implement REDSET will vary between 55 and 179 for all nine schemes. Considering only efficiency, any one of the nine schemes can be used to extract reliability information. For the larger problem with $k = 100$, the total NDFEA required to implement REDSET will vary between 231 and 359 for all nine schemes. These numbers are in hundreds, as expected, and considering the current extremely high computational power, any one of the schemes can be used to extract reliability information.

Obviously, the most attractive scheme needs to be selected considering efficiency, accuracy, and robustness. Using the results shown in Table 5.4 and experiences obtained by solving numerous problems with different degrees of complexity, it will be reasonable to eliminate the experimental Sampling Scheme S1 from further consideration since the accuracy in estimating the underlying risk cannot be assured using it. In the second phase, Sampling Schemes S3, MS3, S3-MLSM, and MS3-IMLSM can also be eliminated since they use the regression analysis to generate an IRS. In the third phase, considering both efficiency and accuracy, Sampling Schemes S2 and MS2 can also be eliminated, since other better options are available. The two remaining Sampling Schemes are S3-KM and MS3-KM. Considering efficiency, accuracy, and robustness, Sampling Scheme MS3-KM appears to be the most attractive. However, before making the final decision, the capabilities of Sampling Scheme MS3-KM to estimate reliabilities of CDNES excited by different types of dynamic loading need to be investigated. In the next four chapters, numerous examples will be considered in an attempt to find the most appropriate sampling design scheme. Using REDSET, underlying risk will be estimated for onshore structures excited by the earthquake loading, offshore structures excited by the earthquake and wave loadings, and solder balls used in computers subjected to thermomechanical loading. The potential of REDSET using Sampling Schemes MS3-KM to implement new design guidelines like the performance-based seismic design (PBSD) concept also needs to be established. If successful, REDSET with Sampling Schemes MS3-KM will provide an alternative to the classical random vibration concept and the basic Monte Carlo simulation technique.

5.10 Development of Reliability Evaluation of Dynamical Engineering Systems Excited in Time Domain (REDSET)

REDSET is a complicated algorithm to estimate the reliability of CNDES. Information needed to initiate the reliability analysis using REDSET is given in Section 5.1. Step-by-step procedures to implement FORM are given in Section 2.10.1.

Considering the complexity of REDSET, it will be helpful if a similar step-by-step implementation of REDSET is also given. These steps are described next. In describing these steps, the original coordinate system is used. All the variables are assumed to be uncorrelated. For correlated variables, all the steps necessary to implement FORM also need to be followed.

5.10.1   Required Steps in the Implementation of REDSET

- Step 1: Identify all the resistance-related RVs and the dynamic loading that need to be considered. Identify all k RVs to generate an IRS of interest. Collect statistical information on them. Select a multiple number of design time histories for the design earthquake, wind-induced wave, or any other dynamic loadings following the procedures discussed in Chapter 4. Procedures to generate the thermomechanical loading will be discussed in Chapter 9. Identify also the LSFs for which the underlying risk needs to be estimated and the corresponding design criterion or the acceptable value.
- Step 2: To iteratively generate an IRS using SD with second-order polynomial without cross terms at the first iteration, assume the coordinates of the initial center point $x_i^*, i = 1, 2, ..., k$, at the mean values of all k RVs.
- Step 3: Excite the structure with the first time history and record the maximum response of interest for the LSF under consideration using any available computer program.
- Step 4: Perform $(2k + 1)$ nonlinear time domain dynamic FE analyses excited by the first time history using SD sampling scheme with second-order polynomial without cross terms. The response data will be used to generate the required IRS in Step 5.
- Step 5: Generate the explicit expression of a second-order IRS without cross terms in terms of vector **b** using Eq. 3.18. Define the corresponding LSF using the information on the permissible or allowable value to formulate the LSF of interest.
- Step 6: Using the LSF generated in Step 5, initiate FORM to estimate the reliability index β. Any value of β, say between 3 and 5 can be assumed; if it is chosen appropriately, the algorithm will converge efficiently in a very few steps. An initial β value of 3.0 is reasonable for the structural reliability estimation for routine applications.
- Step 7: Compute the mean and standard deviation at the checking point of the equivalent normal distribution for those variables that are non-normal. If the RVs are normal, their mean and standard deviation values can be used directly.
- Step 8: Compute partial derivatives of the limit state equation with respect to the design variables at the checking point x_i^*, i.e. $(\partial g/\partial X_i)^*$.

- Step 9: Compute the direction cosine α_{X_i} at the checking point for the X_i RV as:

$$\alpha_{X_i} = \frac{\left(\dfrac{\partial g}{\partial X_i}\right)^* \sigma_{X_i}^N}{\sqrt{\displaystyle\sum_{i=1}^{k}\left(\dfrac{\partial g}{\partial X_i}\sigma_{X_i}^N\right)^{2*}}} \tag{5.25}$$

- Step 10: Compute the new checking point value x_i^* for the X_i RV as:

$$x_i^* = \mu_{X_i}^N - \alpha_{X_i}\beta\,\sigma_{X_i}^N \tag{5.26}$$

- Step 11: Evaluate Eq. 5.26 at the assumed β value in Step 6. Repeat Steps 7 through 9 until the estimates of α_{X_i} converge with a preassigned tolerance level. A tolerance level of 0.005 is common. Once the direction cosines converge, the new checking point can be estimated keeping β in Eq. 5.26 as the unknown parameter. Using the checking point values with β as the unknown parameter for all the RVs and putting them in the LSF, an updated estimate of β will be available. It will be different from the assumed value in Step 6. The assumption of an initial value of β in Step 6 in necessary only for the sake of this step. Otherwise, Step 6 can be omitted. However, when the experience gains in dealing with numerous risk estimation problems, it is observed that this additional assumption of an initial value of β may improve the robustness of the algorithm.
- Step 12: Using the information on the direction cosines, reduce the total number of significant RVs from k to k_R. All other RVs will now be considered deterministic at their mean values.
- Step 13: Determine the coordinates of the next checking point using Eq. 5.26 using the converged direction cosine value α_{X_i} and the updated β value obtained in Step 11 for all k_R RVs. They will be the coordinates of the next center point.
- Step 14: Generate another IRS around the new center point using SD with second-order polynomial without cross terms in terms of k_R RVs.
- Step 15: Redefine the new LSF in terms of k_R RVs.
- Step 16: When all the direction cosine values with respect to the new LSF converge, update the β value using Step 11.
- Step 17: Repeat Steps 7 to 16 until β converges to a predetermined tolerance level. A tolerance level of 0.001 can be used.
- Step 18: Calculate the coordinates of the final center point using the β value obtained in Step 17. The algorithm converges very rapidly after generating at least 3 IRSs; the initial (with k RVs), one intermediate (with k_R RVs), and the final (k_R or m, $m \le k_R$ RVs).

- Step 19: Generate the explicit expression of a second-order IRS with cross terms using the most appropriate sampling design scheme. It will be discussed in more detail later with examples. Most likely, it will be Sampling Scheme MS3-KM.
- Step 20: Estimate the reliability index β and the corresponding probability of failure p_f of CNDES under consideration when excited by the first design time history. TNDFEA required to implement REDSET will range between a few dozen and a few hundred.
- Step 21: Since multiple time histories need to be used to consider uncertainty in the dynamic loadings, repeat Steps 3 to 20 to consider multiple time histories and estimate the corresponding β values. The novel REDSET concept is now fully developed.
- Step 22: Develop a computer program incorporating all the 21 steps discussed here.

5.11 Concluding Remarks

A novel reliability estimation procedure REDSET consisting of 21 steps is presented in this chapter. Implicit response surfaces are presented explicitly in the form of IRSs. At least nine different sampling design schemes are proposed, and the most efficient, accurate, and robust sampling design scheme is identified. Several advanced computational schemes are integrated producing a compounding effect. Since it is a new procedure, it needs to be verified with the help of many informative examples. In the last four chapters, REDSET will be verified, and its implementation potential will be demonstrated. In Chapter 6, REDSET will be verified for the earthquake excitation. It will be verified for the reliability estimation of offshore structures excited by wave loading in Chapter 7. Reliability estimation of offshore structures using REDSET, where sufficient information on the design earthquake time histories is not available, will also be investigated in Chapter 7. A new upcoming design procedure for earthquake loadings, known as PBSD will be introduced in Chapter 8. The implementation potential of PBSD using REDSET will be demonstrated in this chapter. Reliability evaluation of solder balls subjected to the thermomechanical loading caused by heating and cooling using REDSET will be demonstrated in Chapter 9. The underlying risk of CNDES using REDSET can be estimated within a few hours, or one or two days depending on the size of the problem and the type of the excitation.

6

Verification of REDET for Earthquake Loading Applied in the Time Domain

6.1 Introductory Comments

Since REDSET is new and complicated, it needs to be verified using an independent algorithm in terms of its accuracy and efficiency in extracting reliability information for CNDES. The basic Monte Carlo simulation (MCS) technique is used for this purpose. REDSET is verified for the earthquake loading in this chapter. It will be verified for the wave and thermomechanical loadings in Chapters 7 and 9, respectively.

Two different types of examples are presented for the verification purpose in this chapter. In the first two examples, two relatively small three-story steel frames designed by experts, satisfying the post-Northridge design requirements indicating current design practices, are considered to assure that the results obtained are not incidental but reproducible. The smaller size frame makes it easier to estimate the probability of failure, p_f, obtained by all the nine schemes presented in Chapter 5 to generate an IRS in the context of implementing REDSET and then compare results obtained by the basic MCS method for the verification purpose. TNDFEA required for schemes are also tracked to study their efficiency. In the third example, a well-documented case study is considered and the prediction capabilities of different REDSET schemes are studied. A 13-story steel moment frame building that suffered damages of different severity levels in several structural elements during the Northridge earthquake of 1994 is considered. This case study provides a unique opportunity to verify REDSET in a very comprehensive way not only if it can identify the defective elements but also the severity of the defects. At the same time, whether it can also identify defect-free elements or not. Documentations of this type of unique capability will validate REDSET since no other currently available method can extract similar information. The verifications are expected to significantly advance the state of the art in the reliability evaluation of CNDES

After the conclusive verification, REDSET is used to estimate the probability of failure of three buildings designed for the Los Angeles area, by exciting them with

Reliability Evaluation of Dynamic Systems Excited in Time Domain: Alternative to Random Vibration and Simulation, First Edition. Achintya Haldar, Hamoon Azizsoltani, J. Ramon Gaxiola-Camacho, Sayyed Mohsen Vazirizade, and Jungwon Huh.
© 2023 John Wiley & Sons, Inc. Published 2023 by John Wiley & Sons, Inc.

three sets of multiple site- and structure-specific earthquake time histories. It will document many desirable features in the post-Northridge seismic design guidelines and how they can be implemented for reducing the underlying seismic risk for routine applications improving resiliency and sustainability.

6.2 Verification – Example 1: 3-Story Steel Moment Frame with W24 Columns

Eight three-story frames were designed by experts for the Los Angeles area satisfying all the post-Northridge design requirements proposed by the 1997 National Earthquake Hazards Reduction Program (NEHRP) under the sponsorship of the Federal Emergency Management Agency (FEMA-355F 2000). The study was conducted in a partnership of the Structural Engineers Association of California, Applied Technology Council and California Universities for Research in Earthquake Engineering (SAC). They are expected to display some of the improvements proposed to mitigate damages caused by the Northridge earthquake of 1994. These designs are considered benchmark studies and detailed information on them is readily available. These designs are expected to be used in other subsequent studies in the related areas.

There are several purposes for taking this example. Since the frame was designed by experts satisfying all post-Northridge improvements, the study will document if the suggested modifications are reasonable or not. It may also provide information on the risk for different limit state functions (LSFs). The reliability information may help establish the expected or benchmark value for each LSF. It will also document the reliability estimation capabilities of all nine schemes presented in Chapter 5 and may help identify the most attractive sampling design scheme for developing an IRS in the context of implementing REDSET.

6.2.1 Example 1: Accuracy Study of All 9 Schemes

A three-story four-bay steel frame, as shown in Figure 6.1, is considered in Example 1. Member sizes are given in the figure. For this example, all columns are made with W24-sized members.

The frame is represented by FEs. For the reliability evaluation, the statistical characteristics of all the required RVs are collected from the literature for routine applications. The information is summarized in Table 6.1.

The mean value of the modulus of elasticity, E, is considered to be 1.999E8 kN/m^2. The mean values of the yield stress, F_y, for columns and girders are considered to be 4.529E5 kN/m^2 and 3.261E5 kN/m^2, respectively. The mean values of the dead load for floors and roofs are considered to be 33.887 kN/m and

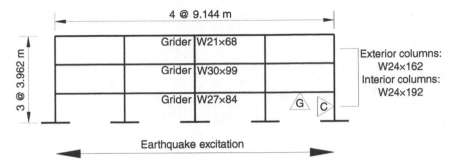

Figure 6.1 Representation of 3-story steel frame.

Table 6.1 Statistical information for random variables.

Variable type	RV	Distribution	COV
(1)	(2)	(3)	(4)
Material properties	Modulus of elasticity (E)	Lognormal	0.06
	Yield stress (F_y)	Lognormal	0.1
Cross-sectional properties	Sectional area (A)	Lognormal	0.05
	Moment of inertia (I)	Lognormal	0.05
	Section modulus (Z)	Lognormal	0.05
Load	Dead load (DL)	Normal	0.1
	Live load (LL)	Type 1	0.25
Earthquake characteristics	Earthquake intensity (g_e)	Type 1	0.1

32.705 kN/m, respectively. The mean value of the live load for floors and roof is considered to be 3.751 kN/m (FEMA-355C 2000). The mean values of the cross-sectional properties of the members are considered to be the values given in the steel design manual (AISC 2011). The uncertainty in terms of coefficient of variation (COV) and the distribution information of all these variables are also given in the table. To incorporate uncertainty in the earthquake intensity, a factor, denoted as g_e, is introduced. Its statistical characteristics are also appropriately defined in Table 6.1.

The frame is excited for 30 seconds by the recorded time history of the N-S component of the 1940 El Centro earthquake, as shown in Figure 6.2.

A girder and a column, indicated in Figure 6.1, are selected and their reliabilities are estimated using REDSET for the strength LSF. To study serviceability LSF, the permissible inter-story drift for the first floor and the overall lateral displacement

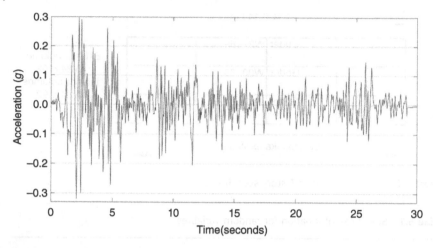

Figure 6.2 N-S component of 1940 El Centro time history.

at the top of the frame are considered not to exceed 0.7% of their respective heights (FEMA-273 1997). Since the maximum displacement is estimated by the nonlinear time history analysis, the permissible values are increased by 125% according to ASCE/SEI 7-10 (ASCE 2010). Therefore, δ_{allow} for inter-story drift and overall lateral displacement are considered 3.47 cm and 10.4 cm, respectively.

The total number of RVs present in the frame for both serviceability and strength limit states is 35. In both cases, 7 RVs are considered to be the reduced number of significant RVs based on their sensitivity indexes. Thus, for this example, $k = 35$ and $k_R = 7$. The reliability index β values and the corresponding probabilities of failure, p_f, for serviceability and strength LSFs are estimated using all nine sampling design schemes presented in Chapter 5, namely Schemes S1, S2, MS2, S3, MS3, S3-MLSM, MS3-MLSM, S3-KM, and MS3-KM. The results are summarized in Table 6.2. Considering the estimated p_f values, Haldar and Mahadevan (2000a) suggested that the number of simulation cycles for the basic MCS should be at least $10 \times (1/p_f)$. The number of simulation cycles is very conservatively considered as 500 000 for this example for the verification purpose. To generate benchmark values for p_f for different LSFs and to compare nine sampling schemes, 500 000 cycles of MCS were conducted. It required about 850 hours of continuous running of a computer. The results using MCS are also summarized in Table 6.2. For comparison, REDSET with Sampling Scheme MS3-KM required about 16 minutes for each run. Assuming TNDFEA of about 100, it will take about 26 hours of computer time to extract reliability information.

From the results summarized in Table 6.2, several important observations can be made. It is observed that the p_f values estimated by all nine schemes are very

Table 6.2 Reliability analysis (verification – 3-story steel moment frame) (W24 columns).

Schemes			S1	S2	MS2	S3	MS3	S3-MLSM	MS3-MLSM	S3-KM	MS3-KM	MCS
(1)			(2)	(3)	(4)	(5)	(6)	(7)	(8)	(9)	(10)	(11)
					AFD			MLSM		KM		
$h^i=1$ $h^f=1$	Serviceability limit state	Inter-story drift	β									
			3.2103	3.2107	3.2106	3.2367	3.2367	3.2370	3.1893	3.2071	3.2071	3.2133
			P_f(E−4) 6.6293	6.6215	6.6240	6.0471	6.0471	6.0402	7.1301	6.7036	6.7036	6.5600
			TNDFEA 101	122	112	229	105	229	105	229	105	500000
		Overall displacement	β 4.1047	4.1049	4.1048	4.1413	4.0928	4.1395	4.0931	4.1059	4.1012	4.0652
			P_f(E−5) 2.0238	2.0225	2.0232	1.7267	2.1308	1.7402	2.1280	2.0137	2.0551	2.4000
			TNDFEA 116	137	127	244	120	244	120	244	120	500000
	Strength limit state	Column C	β 3.4931	3.4931	3.4931	3.5241	3.5239	3.5232	3.5234	3.4930	3.4905	3.4851
			P_f(E−4) 2.3871	2.3870	2.3870	2.1244	2.1266	2.1317	2.1300	2.3883	2.4108	2.4600
			TNDFEA 116	137	127	244	124	244	124	244	120	500000
		Girder G	β 3.1408	3.1411	3.1411	3.1903	3.1903	3.1906	3.1322	3.1413	3.1341	3.1375
			P_f(E−4) 8.4254	8.4164	8.4163	7.1073	7.1073	7.0996	8.6756	8.4110	8.4110	8.5200
			TNDFEA 116	137	127	244	124	244	124	244	120	500000
$h^i=1$ $h^f=0.5$	Serviceability limit state	Inter-story drift	β 3.2126	3.2126	3.2126	3.2093	3.2093	3.2100	3.1993	3.2095	3.2065	3.2133
			P_f(E−4) 6.5774	6.5773	6.5775	6.6534	6.6534	6.6364	6.8885	6.6476	6.7177	6.5600
			TNDFEA 101	122	113	229	105	229	105	229	105	500000
		Overall displacement	β 4.1029	4.1027	4.1028	4.1115	4.1115	4.1107	4.1032	4.1056	4.1052	4.0652
			P_f(E−5) 2.0403	2.0421	2.0407	1.9656	1.9656	1.9722	2.0375	2.0163	2.0201	2.4000
			TNDFEA 116	137	127	244	120	244	120	244	120	500000

(Continued)

Table 6.2 (Continued)

Schemes (1)			AFD					MLSM		KM		MCS
			S1 (2)	S2 (3)	MS2 (4)	S3 (5)	MS3 (6)	S3–MLSM (7)	MS3–MLSM (8)	S3–KM (9)	MS3–KM (10)	(11)
Strength limit state	Column C	β	3.4932	3.4932	3.4932	3.5015	3.5015	3.5011	3.4967	3.4932	3.4929	3.4851
		P_f(E−4)	2.3868	2.3868	2.3868	2.3136	2.3136	2.3169	2.3548	2.3868	2.3891	2.4600
		TNDFEA	116	137	127	244	120	244	120	244	120	500000
	Girder G	β	3.1411	3.1414	3.1414	3.1513	3.1513	3.1513	3.1422	3.1416	3.1406	3.1375
		P_f(E−4)	8.4156	8.4085	8.4084	8.1261	8.1261	8.1277	8.3830	8.4024	8.4299	8.5200
		TNDFEA	116	137	127	244	120	244	120	244	120	500000
$h^i = 1$ $h^f = 0.25$ Serviceability limit state	Inter-story drift	β	3.2126	3.2126	3.2126	3.2163	3.2163	3.2163	3.2096	3.2123	3.2111	3.2133
		P_f(E−4)	6.5773	6.5775	6.5778	6.4917	6.4917	6.4931	6.6464	6.5849	6.6116	6.5600
		TNDFEA	101	122	112	229	105	229	105	229	105	500000
	Overall displacement	β	4.1028	4.1026	4.1028	4.1049	4.1040	4.1042	4.1042	4.1027	4.1041	4.0652
		P_f(E−5)	2.0408	2.0429	2.0413	2.0221	2.0304	2.0285	2.0282	2.0419	2.0298	2.4000
		TNDFEA	116	137	127	244	120	244	120	244	120	500000
Strength limit state	Column C	β	3.4932	3.4932	3.4932	3.4932	3.4932	3.4932	3.4945	3.4932	3.4931	3.4851
		P_f(E−4)	2.3867	2.3867	2.3867	2.3863	2.3863	2.3863	2.3744	2.3867	2.3868	2.4600
		TNDFEA	116	137	127	244	120	244	120	244	120	500000
	Girder G	β	3.1412	3.1414	3.1414	3.1415	3.1411	3.1416	3.1412	3.1414	3.1412	3.1375
		P_f(E−4)	8.4134	8.4066	8.4066	8.4052	8.4144	8.4025	8.4135	8.4067	8.4122	8.5200
		TNDFEA	116	137	127	244	120	244	120	244	120	500000

similar to that of MCS. It indicates that all nine schemes presented are capable of estimating p_f for both LSFs. The results reflect the solid foundation of the REDSET concept; any reasonable design scheme to generate an IRS will produce an acceptable result. All of these schemes extracted reliability information by conducting between 101 and 244 TNDFEA instead of half of a million cycles of MCS. This example validates the proposition that the reliability information of CDNES can be extracted with very few deterministic nonlinear time domain FE analyses. This observation also confirms that the REDSET concept provides alternatives to MCS satisfying one of the major objectives of the study.

As expected, the three-story frame is relatively small and simple and expected to be laterally stiff enough to satisfy the overall lateral drift requirement. The reliability index β is higher and the corresponding p_f is lower for this serviceability LSF confirming the expectations. The study conclusively confirms that the post-Northridge design guidelines are acceptable. Pre- and post-Northridge design guidelines will be compared with examples more elaborately in Chapter 8.

In developing all the schemes, attempts were made to reduce the simulation region as much as possible. The effect of the sampling region size, controlled by the variable h in Eq. 3.10, is studied next. From the results presented so far, h is considered to be 1.0 for all the intermediate and final iterations. To study the effect of the size of the sampling region on the estimated p_f values, h is considered to be 0.5 and 0.25 in the final iteration. The results are also summarized in Table 6.2. It is clear that the reduction of the simulation region improves the estimation of p_f. The regression analysis is needed to generate an IRS when sampling schemes S3, S3-MLSM, and MS3-MLSM are used. The results indicate that they will produce better results for a reduced value of h, say 0.25. For all other schemes based on SD and Kriging, LSFs pass through the experimental sampling points. The results are not highly sensitive to the size of the sampling region for these cases.

As mentioned earlier, the frame was designed by experienced engineers. One of the basic intents of reliability-based designs is that the underlying risks should be similar for different LSFs. The p_f values summarized in Table 6.2 indicate that they are similar for the inter-story drift serviceability LSF and for the strength LSFs of the girders and columns for all nine sampling design schemes. As expected, the 3-story frame is relatively stronger to satisfy the overall deflection requirements.

All nine schemes are accurate in estimating the underlying risk. With the successful verification of all the sampling schemes, it is now necessary to study the efficiency by considering TNDFEA required to implement REDSET. For this example, with $k = 35$, $k_R = 7$, and assuming $n = 3$ and $m = 2$, TNDFEAs required to implement REDSET for all nine schemes are summarized in Table 6.3. As expected, for $k_R = 7$, Scheme S1 required the minimum TNDFEA of 101 and Scheme S3 required the maximum TNDFEA of 229. Schemes MS3, MS3-MLSM, and MS3-KM each will require 105 TNDFEA. The p_f values obtained by Schemes

Table 6.3 Number of deterministic evaluations for different schemes - Example 1 with W24 columns.

(1)	(2)	(3)	(4)	Example 1	
				$k=35, k_R=7, n=3, m=2$	Example 1
				(5)	(6)
First iteration			$2k+1$	71	Example 1
Intermediate iteration			$(n-2)(2k_R+1)$	15	
Final iteration	AFD	S1	$2k_R+1$	15	101
		S2	$(k_R+1)(k_R+2)/2$	36	122
		MS2	$2k_R+1+m(2k_R-m-1)/2$	26	112
		S3	$2^{k_R}+2k_R+1$	143	229
		MS3	2^m+2k_R+1	19	105
	MLSM	S3-MLSM	$2^{k_R}+2k_R+1$	143	229
		MS3-MLSM	2^m+2k_R+1	19	105
	Kriging	S3-KM	$2^{k_R}+2k_R+1$	143	229
		MS3-KM	2^m+2k_R+1	19	105

MS3 and MS3-KM are very similar. However, the p_f values estimated by Scheme MS3-KM need less than half TNDFEA than that of Scheme S3 without compromising the accuracy. Based on the experiences gained dealing with more complex and difficult structures, sampling Scheme S1 is found not to be as accurate as other schemes and can be eliminated from further consideration. Sampling Schemes S3-MLSM and MS3-MLSM require the regression equation in the generation of the final IRS. These two schemes can also be removed from further consideration. The remaining six schemes are promising and need to be evaluated further with the help of more examples.

6.2.2 Verification – Example 2: 3-Story Steel Moment Frame with W14 Columns

As mentioned in Example 1, SAC experts designed 8 three-story buildings in the same Los Angeles area. To confirm all the observations made in Example 1 are valid, another three-story four-bay steel frame designed by them but with column sizes of W14, as shown in Figure 6.2, is considered for this example. Member sizes are shown in the figure. For the reliability evaluation, the statistical information of 8 sensitive reduced number of RVs is summarized in Table 6.4. The frame is then excited by the same time history of the N-S component of the 1940 El Centro earthquake, as shown earlier in Figure 6.2. For the verification purpose, the frame is represented by the same FE's representation as in Example 1.

The first objective of this example is to demonstrate that REDSET is very robust, reliable, and reproducible. It will estimate the underlying risk very accurately for structures designed by experts. The second objective is to develop benchmark information on the acceptable risk for different LSFs that can be specified for future seismic designs. Structures considered in the two examples were designed similarly by the same experts. The third objective is that REDSET can identify the most critical LSF considering both examples.

A girder and a column, indicated in Figure 6.3, are selected to study their reliabilities for the strength LSF. Since the depth of the columns was reduced from W24 to W14, the serviceability LSF of the overall lateral displacement at the top of the frame is considered. The δ_{allow} is considered 10.4 cm, the same value used in Example 1.

For this example, the total numbers of RVs k for the serviceability and strength limit states are 19 and 35, respectively. Eight of them are considered to be the most sensitive reduced number of RVs k_R, for both the serviceability and strength LSFs. The reliability index β and the corresponding probabilities of failure, p_f, values for the serviceability and strength LSFs are estimated by REDSET using seven schemes, namely Sampling Schemes S1, S2, MS2, S3, MS3, S3-KM, and MS3-KM. The results are summarized in Table 6.5. To study the influence of the size

Table 6.4 Statistical information for most important random variables.

Numerical study	Sensi-tivity order	Serviceability		Strength – column		Strength – girder	
		RV	Mean value	RV	Mean value	RV	Mean value
3-story steel frame El-Centro	1	g_e	1	g_e	1	g_e	1
	2	DL^1 (kN/m^2)	33.887	F_{yC} (kN/m^2)	4.529E5	F_{yG} (kN/m^2)	3.261E5
	3	DL^2 (kN/m^2)	33.887	Z_{yC}^{Ext} (m^3)	4.031E−3	Z_{xG}^1 (m^3)	5.113E−3
	4	E (kN/m^2)	1.999E8	DL^2 (kN/m^2)	33.887	DL^2 (kN/m^2)	33.887
	5	I_{xC}^{Int} (m^4)	1.598E−3	I_{xC}^{Ext} (m^4)	1.415E−3	DL^3 (kN/m^2)	32.705
	6	I_{xG}^2 (m^4)	2.456E−3	DL^3 (kN/m^2)	32.705	I_{xG}^1 (m^4)	1.661E−3
	7	I_{xG}^1 (m^4)	1.661E−3	DL^1 (kN/m^2)	33.887	DL^1 (kN/m^2)	33.887
	8	LL^3 (kN/m^2)	3.751	I_{xC}^{int} (m^4)	1.598E−3	I_{xC}^{int} (m^4)	1.598E−3

Superscript stands for floor level, subscript stands for the type of member (C = column, G = girder) in Figure 6.3.
Superscripts Int and Ext stand for interior and exterior columns, respectively.

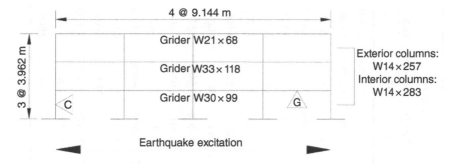

Figure 6.3 Representation of 3-story steel frame, W14.

of the sampling region, h is considered to be 1 and 0.5 in the final iteration. Results for both cases are also shown in Table 6.5. For this example, the total number of the basic MCS cycles is very conservatively considered to be 300 000 for the verification purpose. The results using MCS are also shown in Table 6.5. For this example, it took about 510 hours of continuous running of a computer for a typical MCS.

The major difference between Examples 1 and 2 is the depth of the columns. In Example 1, W24 size of columns are used and in Example 2, W14 size of columns are used. The frame in Example 1 is much stronger in the lateral direction than the frame in Example 2. The p_f value for the overall lateral deflection increased significantly for Example 2 with W14 columns. The estimated p_f values are similar for the two strength LSFs in both examples. The p_f values obtained by all seven schemes using REDSET and 300 000 cycles of MCS are very similar indicating that they are viable and accurate. All other observations made for Example 1 are also valid for this example. The results also indicate that Scheme S1 will require a minimum TNDFEA of only 73 to implement REDSET for the overall displacement LSF. For this small problem, the accuracy of Scheme S1 is similar to other schemes. However, the accuracy cannot be assured for other complicated structures. For this example, it will require TNDFEA of 105 when Sampling Scheme MS3-KM is used instead of 329 when Sampling Scheme S3-KM is used. Obviously, REDSET with sampling scheme MS3-KM will require a smaller number of TNDFEA compared to other similar schemes. Similar observations can also be made by considering other examples. In general, REDSET with sampling scheme MS3-KM appears to be the best option for different structural systems considering both accuracy and efficiency.

As stated earlier, both frames were designed by experienced professionals satisfying all the requirements that existed in 1997. The results indicate that the p_f values are similar for the serviceability and strength LSFs. One of the major intentions of the risk-based design concept is to make p_f values similar for different LSFs. This observation is very exciting and observing that the post-Northridge

Table 6.5 Reliability analysis (verification−3-story steel moment frame) (W14 columns).

Schemes			AFD					Kriging		
			S1	S2	MS2	S3	MS3	S3-KM	MS3-KM	MCS
(1)			(2)	(3)	(4)	(5)	(6)	(9)	(10)	(11)
$h^i=1$ $h^f=1$	Overall displacement	β	3.5521	3.5521	3.5520	3.5348	3.5375	3.5441	3.5439	3.5583
		PF (E−4)	1.911	1.911	1.912	2.040	2.020	1.970	1.9710	1.866
		TNDFEA	73	101	86	329	81	329	105	3E05
	Strength limit state Column C	β	3.4662	3.4663	3.4663	3.5070	3.4612	3.4663	3.4621	3.4471
		PF (E−4)	2.639	2.638	2.638	2.266	2.689	2.638	2.680	2.833
		TNDFEA	139	167	152	395	147	395	147	3E05
	Girder G	β	3.9279	3.9285	3.9285	3.9510	3.9579	3.9282	3.9204	3.9073
		PF (E−5)	4.285	4.274	4.274	3.891	3.781	4.279	4.420	4.666
		TNDFEA	139	167	152	395	155	395	147	3E05
$h^i=1$ $h^f=0.5$	Overall displacement	β	3.5516	3.5511	3.5512	3.5832	3.5773	3.5530	3.5484	3.5583
		PF (E−4)	1.915	1.918	1.917	1.697	1.736	1.904	1.938	1.866
		TNDFEA	73	101	86	329	81	329	77	3E05
	Strength limit state Column C	β	3.4659	3.4661	3.4661	3.4745	3.4807	3.4663	3.4666	3.4471
		PF (E−4)	2.642	2.640	2.640	2.559	2.501	2.638	2.635	2.833
		TNDFEA	139	167	152	395	143	395	143	3E05
	Girder G	β	3.9274	3.9280	3.9280	3.9246	3.9202	3.9275	3.9280	3.9073
		PF (E−5)	4.294	4.283	4.283	4.344	4.424	4.292	4.283	4.666
		TNDFEA	139	167	152	395	143	395	143	3E05

h^i is the value used for intermediate iterations
h^f is the value used for the final iteration

seismic design guidelines are moving in the right direction. This study confirms the appropriateness of the advanced concepts incorporated in the current codes. Based on this example, the overall accepted seismic risk appears to be between 10^{-5} and 10^{-4}. This type of acceptable risk was also observed for steel structures using the Load and Resistance Factor Design (LRFD) concept commonly used at present. The information on this acceptable risk can help design other seismic risk-tolerant structures. Another important observation is that the selection of simulation regions, in terms of h is an important parameter that influences the reliability information. However, it may not be a deciding factor when KM is used. This simple example again validates all seven methods considered earlier in Example 1.

Both examples clearly indicate that the reliability evaluation of CDNES using REDSET significantly advanced the state of the art. The underlying risk can be estimated with few hundreds of deterministic FE analyses. The exact NDFEA will depend on the values of k, k_R, m, and n. The efficiency can be significantly improved without compromising accuracy without considering cross terms for all k_R RVs, i.e. considering $k_R = m$. Of course, when m is smaller than k_R, the efficiency will be increased accordingly. However, they are not excessive but implementable, particularly considering existing enormous computational capabilities.

After verifying REDSET with two examples, it will be desirable to take a well-documented case study to conclusively confirm its other unique capabilities.

Before moving forward, it will be desirable to expand on our discussions on selecting the most attractive or optimum sampling scheme. Out of the seven remaining schemes, Sampling Scheme S1 was already eliminated. Sampling Scheme S2 may not satisfy the accuracy requirement. Sampling Scheme S3 uses regression analysis to generate the last IRS. Sampling Scheme S3-KM can be easily replaced with Sampling Scheme MS3-KM. Considering both efficiency and accuracy, three remaining modified sampling schemes, namely MS2, MS3, and MS3-KM appear to be reasonable. They are investigated further using a case study, as discussed next.

6.3 Case Study: 13-Story Steel Moment Frame

After the successful verification of REDSET with different experimental sampling design schemes for two frames designed by experts, the stage is now set to document few unique capabilities of it using a well-documented case study using the three modified schemes mentioned above.

An actual 13-story steel moment frame building located in the Southern San Fernando Valley in California, north of Route 101 is considered. The building suffered significant damages during the Northridge earthquake of 1994 and all

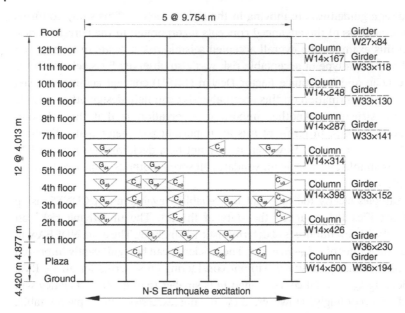

Figure 6.4 Representation of 13-story steel frame.

necessary information required for a case study is well documented in the literature (Uang et al. 1995). The inspection program was sponsored by the National
Science Foundation. Observations made during the inspection are very detailed
and expected to be accurate. The inspection report is widely available. Information
provided in the report is used for this case study. The building consisted
of steel frames as shown in Figure 6.4. Obviously, the frame is not expected to satisfy all the suggested post-Northridge design improvements and requirements
expected at present.

Detailed information on the severity of damages to the members caused by the
earthquake is reported in the inspection report. Seven girders (G_{d1} to G_{d7}) and
seven columns (C_{d1} to C_{d7}), identified in Figure 6.4, suffered the most significant
amount of damages (severely damaged). Near these elements, three girders (G_{m1} to
G_{m3}) and three columns (C_{m1} to C_{m3}) suffered moderate amount of damages (moderately damaged). Three columns (C_{u1} to C_{u3}) and three girders (G_{u1} to G_{u3}) did
not suffer any damage but close to the severely damaged columns are also selected
to check the capabilities of REDSET. The major objective is to study whether the p_f
values obtained by REDSET can be correlated with the corresponding damaged
states and severity levels of damage. Obviously, p_f values will not be zero for
undamaged members, but it will be interesting to investigate whether they fall
below the acceptable risk value or not. If successful, the case study will document
some of the major capabilities of REDSET.

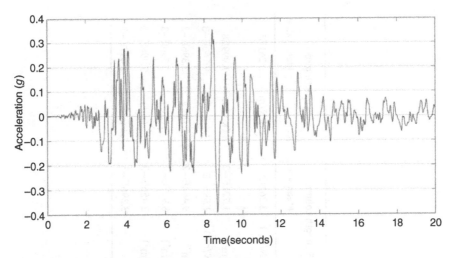

Figure 6.5 Northridge time history at Canoga Park station.

Numerous earthquake time histories were recorded in the vicinity of the damaged building. The frame is excited by the earthquake time history closest to the building exhibiting similar soil characteristics. It was recorded at the Canoga Park station and its time history is shown in Figure 6.5.

The sizes of the members are shown in Figure 6.4. The mean values of the material properties, E and F_y are considered to be 1.999E8 kN/m^2 and 3.261E5 kN/m^2, respectively. The mean values of the gravitational load acting on typical floors and roof are reported to be 40.454 kN/m and 30.647 kN/m, respectively, at the time of the earthquake. Statistical information on the eight most significant reduced RVs is summarized in Table 6.6.

The frame is represented by FEs. The columns and girders are represented by beam-column elements. Their probabilities of failure are estimated for the strength LSF represented by the interaction equations as in Eqs. 3.25 and 3.26. To study the serviceability LSF, the inter-story drift represented by Eq. 3.20 is used. The permissible inter-story drift is considered not to exceed 0.7% of their respective height. Similar to the previous examples, the permissible value was increased by 125% according to the ASCE/SEI 7-10 (2010).

For the strength LSFs of columns and girders, the total number of RVs k are estimated to be 157 and 101, respectively. For the serviceability LSF, $k = 72$. After conducting the sensitivity analysis, the most sensitive RVs, k_R is found to be 8 for both the strength and serviceability LSFs. The p_f values of all the columns highlighted in Figure 6.4 are estimated using REDSET with modified Sampling Schemes MS2, MS3, and MS3-KM. The results are summarized in Table 6.7 for the highlighted severely and moderately damaged columns and a few undamaged

Table 6.6 Statistical information for most important reduced random variables.

Numerical study	Sensitivity order	Serviceability		Strength column		Strength girder	
		RV	Mean value	RV	Mean value	RV	Mean value
13-story steel frame Canoga Park station	1	g_e	1	F_y (kN/m^2)	3.261E5	F_y (kN/m^2)	3.261E5
	2	E (kN/m^2)	1.999E8	g_e	1	g_e	1
	3	I_{xC}^{12} (m^4)	8.907E−4	Z_{xC}^{1} (m^3)	1.721E−2	E (kN/m^2)	1.999E8
	4	I_{xG}^{11} (m^4)	2.456E−3	E (kN/m^2)	1.999E8	Z_{xG}^{4} (m^3)	9.160E−3
	5	I_{xC}^{11} (m^4)	1.253E−3	A_{C}^{1} (m^2)	9.484E−2	I_{xG}^{4} (m^4)	3.396E−3
	6	I_{xG}^{12} (m^4)	2.456E−3	I_{xC}^{12} (m^4)	8.907E−4	I_{xG}^{5} (m^4)	3.396E−3
	7	I_{xC}^{13} (m^4)	8.907E−4	I_{xG}^{11} (m^4)	2.456E−3	I_{xG}^{3} (m^4)	3.396E−3
	8	I_{xC}^{1} (m^4)	3.417E−3	(m^4)	3.396E−3	I_{xC}^{5} (m^4)	2.497E−3

Table 6.7 Reliability analysis for the strength LSF of columns (case study – 13-story building).

Schemes			AFD		Kriging	
			MS2	MS3	MS3-KM	MCS
(1)			(2)	(3)	(4)	(5)
Highly damaged elements	C_{d1}	β	1.0771	1.0629	1.0777	1.0762
		p_f (E−1)	1.4071	1.4392	1.4059	1.4092
		TNDFEA	356	349	349	500 000
	C_{d2}	β	1.8146	1.8574	1.8347	1.8279
		p_f (E−2)	3.4796	3.1625	3.3272	3.3780
		TNDFEA	371	364	364	500 000
	C_{d3}	β	1.3413	1.3166	1.3341	1.3279
		p_f (E−2)	8.9919	9.3985	9.1078	9.2102
		TNDFEA	356	349	349	500 000
	C_{d4}	β	1.8539	1.8673	1.8551	1.8526
		p_f (E−2)	3.1876	3.0927	3.1790	3.1968
		TNDFEA	371	364	364	500 000
	C_{d5}	β	1.3433	1.3200	1.3369	1.3413
		p_f (E−2)	8.9587	9.3422	9.0630	8.9918
		TNDFEA	356	349	349	500 000
	C_{d6}	β	1.5832	1.6989	1.7357	1.7264
		p_f (E−2)	5.6682	4.4673	4.1309	4.2134
		TNDFEA	381	488	424	500 000
	C_{d7}	β	1.0062	0.9983	1.0077	1.0064
		p_f (E−1)	1.5717	1.5908	1.5681	1.5710
		TNDFEA	356	349	349	500 000
Moderately damaged elements	C_{m1}	β	2.6226	2.6212	2.6293	2.6246
		p_f (E−3)	4.3632	4.3808	4.2779	4.3380
		TNDFEA	386	379	379	500 000
	C_{m2}	β	1.5825	1.6757	1.6780	1.6626
		p_f (E−2)	5.6773	4.6895	4.6678	4.8200
		TNDFEA	371	364	364	500 000
	C_{m3}	β	2.4851	2.4814	2.4880	2.4849
		p_f (E−3)	6.4760	6.5441	6.4227	6.4800
		TNDFEA	371	364	364	500 000

(Continued)

Table 6.7 (Continued)

Schemes			AFD		Kriging	
			MS2	MS3	MS3-KM	MCS
(1)			(2)	(3)	(4)	(5)
No damaged elements	C_{u1}	β	3.6157	3.6156	3.6189	3.5921
		p_f (E−4)	1.4976	1.4982	1.4795	1.6400
		TNDFEA	386	379	379	500 000
	C_{u2}	β	3.4653	3.4818	3.4657	3.4408
		p_f (E−4)	2.6484	2.4902	2.6441	2.9000
		TNDFEA	441	424	452	500 000
	C_{u3}	β	4.1419	4.1453	4.1430	4.1075
		p_f (E−5)	1.7222	1.6965	1.7136	2.0000
		TNDFEA	401	394	394	500 000

The parameter $h = 1$ for intermediate iterations
The parameter $h = 0.25$ for the final iteration

columns as discussed earlier. Similar information on p_f for the highlighted girders is summarized in Table 6.8. The p_f values for the inter-story drift LSF between the 1st floor and Plaza and between the 7th and 6th floors are estimated and the results are summarized in Table 6.9. To validate the estimated risks of failure of the highlighted members using REDSET with three experimental sampling schemes, 500 000 cycles of the basic MCS requiring about 1461 hours of continuous running of a computer were carried out. The results are also shown in the corresponding tables.

Several important observations can be made from the results shown in these tables. The p_f values are very similar to the numbers obtained by MCS indicating that they are accurate. The estimated p_f values correlate extremely well with the levels of damage (severe, moderate, and undamaged). The results are very encouraging and document that estimated p_f values can be correlated with different damage states and damage levels. The results also establish the robustness of REDSET in estimating p_f values for high, moderate, and undamaged states.

The p_f values of the highly and moderately damaged columns and girders are very high; not in the range currently practiced in the profession as demonstrated in Examples 1 and 2. The negative value of reliability index β indicates that p_f is more than 50%. The β values for the inter-story drift between the 7th and 6th floors are negative indicating the p_f values are more than 50%. They will not satisfy any inter-story drift requirement in any design guidelines including the

Table 6.8 Reliability analysis for the strength LSF of girders (case study – 13-story building).

Schemes			AFD		Kriging	MCS
			MS2	MS3	MS3-KM	
(1)			(2)	(3)	(5)	(6)
Highly damaged elements	G_{d1}	β	−0.2195	−0.2091	−0.2213	−0.2227
		p_f	0.5869	0.5828	0.5876	0.5881
		TNDFEA	244	237	237	500000
	G_{d2}	β	−0.0387	−0.0095	−0.0388	−0.0418
		p_f	0.5154	0.5038	0.5155	0.5167
		TNDFEA	229	222	222	500000
	G_{d3}	β	0.8607	0.8610	0.8607	0.8593
		p_f	0.1947	0.1946	0.1947	0.1951
		TNDFEA	244	237	237	500000
	G_{d4}	β	1.1223	1.1179	1.1307	1.1289
		p_f	0.1309	0.1318	0.1291	0.1295
		TNDFEA	244	237	237	500000
	G_{d5}	β	−0.7292	−0.8492	−0.8554	−0.8033
		p_f	0.7671	0.8021	0.8038	0.7891
		TNDFEA	244	237	237	500000
	G_{d6}	β	−0.2747	−0.2300	−0.3055	−0.2743
		p_f	0.6082	0.5910	0.6200	0.6081
		TNDFEA	244	249	237	500000
	G_{d7}	β	1.0743	1.0625	1.0738	1.0719
		p_f	0.1413	0.1440	0.1415	0.1419
		TNDFEA	244	237	237	500000
Moderately damaged elements	G_{m1}	β	1.5326	1.5676	1.5564	1.5433
		p_f (E−2)	6.2688	5.8482	5.9802	6.1377
		TNDFEA	289	282	282	500000
	G_{m2}	β	1.4704	1.4708	1.4718	1.4694
		p_f (E−2)	7.0731	7.0677	7.0535	7.0864
		TNDFEA	244	237	237	500000
	G_{m3}	β	1.4722	1.4738	1.4734	1.4713
		p_f (E−2)	7.0484	7.0263	7.0315	7.0604
		TNDFEA	244	237	237	500000

(Continued)

Table 6.8 (Continued)

Schemes			AFD		Kriging	
			MS2	MS3	MS3-KM	MCS
(1)			(2)	(3)	(5)	(6)
No damaged elements	G_{u1}	β	1.9757	1.8817	1.8905	1.9002
		p_f (E−2)	2.4092	2.9940	2.9343	2.8706
		TNDFEA	259	376	376	500000
	G_{u2}	β	1.9946	1.9086	1.9143	1.9036
		p_f (E−2)	2.3042	2.8155	2.7788	2.8483
		TNDFEA	278	391	391	500000
	G_{u3}	β	1.9708	1.8796	1.8896	1.8824
		p_f (E−2)	2.4375	3.0079	2.9404	2.9891
		TNDFEA	259	376	376	500000

The parameter $h = 1$ for intermediate iterations
The parameter $h = 0.25$ for the final iteration

Table 6.9 Reliability analysis inter-story drift LSF (case study – 13-story building).

Schemes		AFD		Kriging	
		MS2	MS3	MS3-KM	MCS
(1)		(2)	(3)	(5)	(6)
Plaza – 1st floor	β	0.5311	0.5280	0.5327	0.5316
	p_f	0.2977	0.2987	0.2971	0.2975
	TNDFEA	186	179	179	500000
6th–7th floor	β	−0.0911	−0.0681	−0.0681	−0.0715
	p_f	0.5363	0.5272	0.5272	0.5285
	TNDFEA	186	183	183	500000

The parameter $h = 1$ for intermediate iterations
The parameter $h = 0.25$ for the final iteration

pre-Northridge design guidelines. The β values for the inter-story drift between the Plaza and 1st floor are not negative but around 0.5 with p_f of 0.3. This is also very high and will not be permissible by any design guidelines. It appears that the inter-story drift LSF was not checked at the design stage. One can argue that the building was designed by satisfying all the pre-Northridge design criteria that existed in the Southern San Fernando Valley area and the inter-story drift was not required to be checked at that time. REDSET clearly identified this major weakness in the pre-Northridge design criteria. Obviously, this weakness was eliminated in developing the post-Northridge design criteria.

Another interesting observation can be made by comparing p_f values given in Tables 6.7 and 6.8 for highly and moderately damaged elements. One can observe that the p_f values of all girders are much higher than that of columns. It reflects the strong column and weak girder concept was used in designing the building, generally practiced before the pre-Northridge time period. It prevented the formation of plastic hinges in the columns. However, very high p_f values of the damaged girders are also not acceptable. According to the current design practices, the acceptable p_f value of the girders and columns in strength is expected to be around 10^{-4}–10^{-6}. Similar values were observed for Examples 1 and 2. Assuming that the procedure used to design girders was reasonable before the pre-Northridge period, one can postulate that the frame was not analyzed properly considering the nonlinear behavior. The study also conclusively confirms that the post-Northridge design criteria are superior to the pre-Northridge design criteria.

The information on the reliability index for the strength LSF of columns and girders for severe, moderate, and undamaged states estimated by REDSET with Sampling Scheme MS3-KM is summarized in Tables 6.7 and 6.8, which is so unique and informative; the results are plotted in Figure 6.6. The plots clearly indicate distinct separation between different damage states and show that the β values increase from severe to no damage states in a very systematic way for both columns and girders. These observations are very exciting and promising to demonstrate several attractive capabilities of REDSET. In fact, there may not be any other reliability evaluation procedure currently available to demonstrate these capabilities.

It will be informative to discuss the relative merits of REDSET with Schemes MS2, MS3, and MS3-MK by comparing the information on p_f obtained in Examples 1 and 2, and in this case study. The stage is now set to comment on the most appropriate experimental sampling scheme for the generation of an IRS in the implementation process of REDSET. The p_f values obtained by the three modified schemes are very similar to the values obtained by MCS indicating that they are acceptable. Results in Tables 6.2, 6.5, and 6.7–6.9 also indicate that REDSET with Scheme MS2 will require a few extra overall TNDFEA in some cases. This observation indicates that Sampling Scheme MS2 may not be the most attractive and can be eliminated from further consideration. The remaining two schemes are Scheme MS3 (an IRS is generated using the regression analysis) and MS3-KM (an IRS is generated using Kriging). By using the process of elimination, REDSET with MS3-KM appears to be the most appropriate reliability evaluation method for CDNESs.

(a)

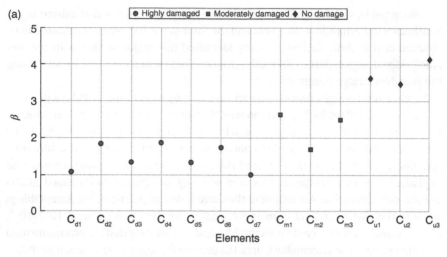

(b)

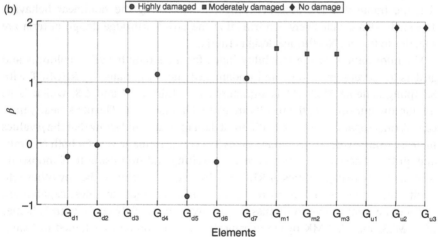

Figure 6.6 Graphical representation of reliability index β for 13-story building using MKM; (a) Column strength; and (b) Girder strength.

6.4 Example 4: Site-Specific Seismic Safety Assessment of CDNES

Unique and extremely desirable capabilities of REDSET are presented to estimate the risk of CNDES excited in the time domain with the help of two examples and a case study. For the verification purpose, structures were excited by one earthquake time history. To incorporate the presence of a considerable amount of uncertainty in the design earthquake time history at a site, current design procedures require considering multiple time histories fitting the design spectrum. It is now necessary to document how to estimate the seismic risk of a structure at a specific site and a structure satisfying the current design requirements. The intent of this example is to showcase the implementation of REDSET for the seismic risk assessment of a structure at the design stage. The steps identified below can be used for a routine evaluation. Information required to initiate the evaluation includes (i) the location where the structure will be built, (ii) the dynamic characteristic of the structure, (iii) uncertainty quantifications of the resistance and load-related design variables, (iv) the target probabilistic ground motion response spectrum (TPGRS), (v) soil condition at the top 30 m of the site, (vi) LSFs required to be satisfied, and (vii) of course, a computer program capable of conducting nonlinear FE-based dynamic analysis excited in the time domain.

Two major objectives of this example are to generate multiple site-specific earthquake time histories and then demonstrate how to estimate the underlying risk using REDSET. With the help of examples, it is demonstrated how to satisfy the intent of the current design guidelines and to document the design process for review.

6.4.1 Location, Soil Condition, and Structures

Three structures already designed by experts are considered to demonstrate how design guidelines can be satisfied using REDSET. A joint venture known as SAC consisting of the Structural Engineers Association of California (SEAOC), the Applied Technology Council (ATC), and California Universities for Research in Earthquake Engineering (CUREe) was formed to design a few buildings satisfying all post-Northridge requirements. SAC designed eight 3-story buildings, eight 9-story buildings, and four 20-story buildings using the 1997 NEHRP provisions in the Los Angeles area. FEMA published these designs so that they can be used for further research in related areas. Since their dynamic characteristics are very different, one design from each of the three categories (3-story, 9-story, and 20-story) is selected for this demonstrative study. They are located in the Los Angeles area at a site with coordinates of 118.243° W and 34.054° N. In the United States, soil types are classified into six categories (Site Classes A to F) based on the upper 30 m of the site profile (ASCE 7-10). Site Class A represents hard rock with measured shear wave velocity greater than about 1500 m/s. Site Class E represents soft clay soil with a measured shear wave velocity of less than about 180 m/s. When the soil properties are not known in sufficient detail to determine the site class, Site Class D must be used. Site Class D represents stiff soil with measured shear wave velocity between about 180 and 360 m/s. Since no information on the soil condition is known for the site, it is considered to be Site Class D.

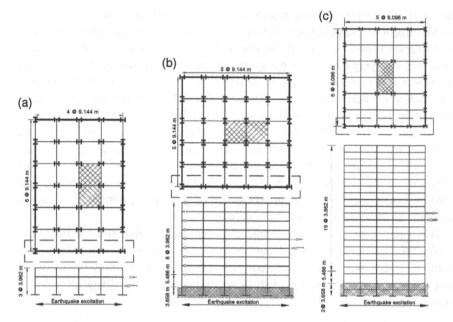

Figure 6.7 Plan and elevation view of three steel buildings. (a) 3-story (b) 9-story (c) 20-story.

The floor plans and elevation for the buildings are shown in Figure 6.7. These were considered to be office buildings and were designed by expert professionals satisfying all applicable design guidelines. Four different column sizes (W14, W24, W30, and W36) were used in each group (FEMA 355F). Results for the buildings designed using W24 columns are reported in this work. Each building is represented by a frame as shown in the figure. Member sizes for these buildings are summarized in Table 6.10. The fundamental periods of these structures are 0.86 seconds, 2.18 seconds, and 3.43 seconds, respectively.

6.4.2 Uncertainty Quantifications

For proper documentation, uncertainties associated with the resistance- and load-related random variables are presented separately.

6.4.3 Uncertainty Quantifications in Resistance-Related Design Variables

Statistical information on all the resistance-related RVs is collected from the literature. The information is summarized in Table 6.11. The uncertainties are described in terms of the ratio of mean to nominal values, nominal and mean values separately, COV, and the corresponding distributions.

Table 6.10 Member sizes for the 3-, 9-, and 20-story buildings.

3-story Columns Exterior	3-story Columns Interior	3-story Girders	9-story Columns Exterior	9-story Columns Interior	9-story Girders	20-story Columns Exterior	20-story Columns Interior	20-story Girders
W24×192	W24×207	W24×76	W24×207	W24×250	W21×62	15 × 15 × 0.5	W24×207	W18×46
W24×192	W24×207	W33×118	W24×207	W24×250	W27×94	15 × 15 × 0.5	W24×207	W24×55
W24×192	W24×207	W30×108	W24×250	W24×279	W33×118	15 × 15 × 0.75	W24×250	W27×84
			W24×250	W24×279	W33×118	15 × 15 × 0.75	W24×250	W30×108
			W24×279	W24×335	W36×150	15 × 15 × 1.0	W24×279	W30×108
			W24×279	W24×335	W36×150	15 × 15 × 1.0	W24×279	W33×118
			W24×335	W24×335	W36×150	15 × 15 × 1.0	W24×279	W33×118
			W24×335	W24×335	W36×182	15 × 15 × 1.0	W24×335	W33×118
			W24×335	W24×408	W36×182	15 × 15 × 1.0	W24×335	W36×135
			W24×335	W24×408	W36×182	15 × 15 × 1.0	W24×335	W36×135
						15 × 15 × 1.25	W24×408	W36×135
						15 × 15 × 1.25	W24×408	W36×135
						15 × 15 × 1.25	W24×408	W36×135
						15 × 15 × 1.25	W24×408	W36×135
						15 × 15 × 1.25	W24×408	W36×135
						15 × 15 × 1.25	W24×408	W36×135
						15 × 15 × 2.0	W24×492	W36×135
						15 × 15 × 2.0	W24×492	W33×118
						15 × 15 × 2.0	W24×492	W33×118
						15 × 15 × 2.0	W24×492	W33×118
						15 × 15 × 2.0	W24×492	W33×118
						15 × 15 × 2.0	W24×492	W14×22

Table 6.11 Statistical information of resistance-related random variables.

RV clasification		RV	$\bar{X}/X_n{}^a$	Nominal (X_n)	Mean ($\bar{X}$)	COV	Dist.
Material properties	Column	E (kN/m²)	1.00	1.9995E8	1.9995E+8	0.06	Lognormal
		F_y (kN/m²)	1.15	344738	396449	0.1	Lognormal
		b ^{b}	1.00	0.02	0.02	0.1	Lognormal
	Girder	E (kN/m²)	1.00	1.9995E8	1.9995E8	0.06	Lognormal
		F_y (kN/m²)	1.35	248211	335085	0.1	Lognormal
		b	1.00	0.02	0.02	0.1	Lognormal
Sample of cross-sectional properties		A (m²)	1.00	From AISC steel construction manual		0.05	Lognormal
		I_x (m⁴)	1.00			0.05	Lognormal
		I_y (m⁴)	1.00			0.05	Lognormal
		Z_x (m³)	1.00			0.05	Lognormal
		Z_y (m³)	1.00			0.05	Lognormal
		S_x (m³)	1.00			0.05	Lognormal
		S_y (m³)	1.00			0.05	Lognormal

^{a} $\bar{X}/X_n$ stands for mean to nominal ratio.
^{b} b stands for the ratio between post-yield tangent and initial elastic tangent (strain-hardening ratio).

6.4.3.1 Uncertainty Quantifications in Gravity Load-related Design Variables

The gravity loads include dead load acting at different locations and live load as reported by SAC. The information is summarized in Table 6.12.

6.4.3.2 Selection of a Suite of Site-Specific Acceleration Time Histories

Assuming Soil Class D, the TPGRS for the site is developed, as shown in Figure 6.8, according to ASCE 7-10. For this example, using Alternative 1 discussed in

Table 6.12 Statistical information of gravity load-related random variables.

RV classification		RV	$\overline{X}/X_n{}^a$	Nominal	Mean	COV	Dist.
Load	Gravity loads	DL_F (kN/m²)	1.05	4.5965	4.3236	0.1	Normal
		DL_R (kN/m²)	1.05	3.9741	4.1728	0.1	Normal
		DL_P (kN/m²)	1.05	5.5541	5.8318	0.1	Normal
		DL_W (kN/m²)	1.05	1.1970	1.2569	0.1	Normal
		$LL_{F\&R}$ (kN/m²)^b	0.50	0.9576	0.4788	0.25	Type 1

Subscripts F, R, P, and W stand for Floor level, Roof level, Penthouse, and Wall Surface, respectively.

^{a} $\overline{X}/X_n$ stands for mean to nominal ratio.
^{b} Reduced LL.

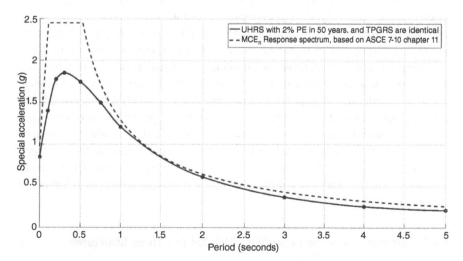

Figure 6.8 Los Angeles with site class D-category soil with average shear velocity of 259 m/s at a depth of 30 m.

Chapter 4, 11 acceleration time histories are selected by scaling past recorded data available at Pacific Earthquake Engineering Research Center (PEER) database. For the Los Angeles site under consideration, $C_R = 1.0$ and TPGRS and UHRS will be identical. The site-specific TPGRS is represented by UHRS in Figure 6.8. The Risk-Targeted Maximum Considered (MCE_R) response spectrum is also shown in the figure.

With the availability of TPGRS and following the procedures discussed in Chapter 4, Section 4.3, 11 acceleration time histories for each structure are selected. They are listed in Table 6.13 with their original identification nomenclature used by PEER with the scale factors used to fit the TPGRS. Actually recorded acceleration time histories of these earthquakes are shown in Azizsoltani (2017). The spectral accelerations of the selected 11 ground motions using TPGRS are matched at the respective fundamental periods of the structures.

6.5 Risk Evaluation of Three Structures using REDSET

6.5.1 Selection of Limit State Functions

The overall lateral displacement at the top of the frames and the inter-story drift are the two serviceability LSFs considered in this example. To define these serviceability LSFs, the information on the allowable deflection δ_{allow} needs to be known. Three buildings considered in this example were designed according to the post-Northridge guidelines. They are expected to satisfy the performance-based seismic design (PBSD) requirements. PBSD advocates for designing structures for different performance levels with a specified underlying risk that users/owners are willing to take replacing the life safety (LS) criterion used in the past. Performance levels are defined as immediate occupancy (IO) (a performance level of very low structural damage), LS, and collapse prevention (CP) (a state of extreme structural damage). PBSD suggested δ_{allow} values for the overall lateral displacement and inter-story drift for the three performance levels. δ_{allow} values suggested for the CP performance level for the three buildings are used to develop the required serviceable LSFs. For the overall lateral displacement, LSFs are considered to be 0.743, 2.324, and 5.048 m, respectively for the three buildings. For the inter-story drift LSF, δ_{allow} is considered to be 0.248 m for all three buildings.

6.5.2 Estimations of the Underlying Risk for the Three Structures

Necessary information to estimate the reliability index β and the corresponding p_f using REDSET with Sampling Scheme MS3-KM for the three buildings is now

Table 6.13 Selected time histories for 3-, 9-, and 20-story structures.

	Record	Record name	Scale factor	PEER description
3-story	1	CHICHI/TCU055-N.at2	3.93	Chi-Chi Taiwan, 9/20/1999, TCU055, N
	2	NORTHR/RRS318.at2	1.54	Northridge-01, 1/17/1994, Rinaldi Receiving Sta, 318
	3	SUPERST/B-IVW360.at2	2.97	Superstition Hills-02, 11/24/1987, Imperial Valley, 360
	4	NORTHR/NWH360.at2	1.09	Northridge-01, 1/17/1994, Newhall - Fire Sta, 360
	5	CHICHI/TCU049-N.at2	3.03	Chi-Chi Taiwan, 9/20/1999, TCU049, N
	6	NORTHR/LDM334.at2	2.11	Northridge-01, 1/17/1994, LA Dam, 334
	7	COYOTELK/G03140.at2	3.58	Coyote Lake, 8/6/1979, Gilroy Array #3, 140
	8	CHICHI/TCU101-N.at2	3.73	Chi-Chi Taiwan, 9/20/1999, TCU101, N
	9	CHICHI/TCU067-N.at2	1.83	Chi-Chi Taiwan, 9/20/1999, TCU067, N
	10	CHICHI/TCU042-N.at2	3.90	Chi-Chi Taiwan, 9/20/1999, TCU042, N
	11	NORTHR/STC180.at2	1.63	NORTHRIDGE EQ 1/17/94, 12:31, USC STATION 90003
9-story	1	CHICHI/TCU089-E.at2	3.38	Chi-Chi Taiwan, 9/20/1999, TCU089, E
	2	CHICHI/ILA013-W.at2	3.18	Chi-Chi Taiwan, 9/20/1999, ILA013, W
	3	CHICHI/TCU065-N.at2	1.22	Chi-Chi Taiwan, 9/20/1999, TCU065, N
	4	IMPVALL/H-E08140.at2	2.24	Imperial Valley-06, 10/15/1979, El Centro Array #8, 140
	5	LOMAP/HSP090.at2	3.61	Loma Prieta, 10/18/1989, Hollister - South & Pine, 90
	6	LOMAP/HCH180.at2	2.12	Loma Prieta, 10/18/1989, Hollister City Hall, 180
	7	SUPERST/B-POE360.at2	3.18	Superstition Hills-02, 11/24/1987, Poe Road (temp), 360

(*Continued*)

Table 6.13 (Continued)

	Record	Record name	Scale factor	PEER description
	8	LOMAP/SLC270.at2	3.64	Loma Prieta, 10/18/1989, Palo Alto - SLAC Lab, 270
	9	ERZIKAN/ERZ-EW.at2	1.59	Erzican Turkey, 3/13/1992, Erzincan, EW
	10	CHICHI/TCU054-N.at2	3.28	Chi-Chi Taiwan, 9/20/1999, TCU054, N
	11	IMPVALL/I-ELC270.at2	3.10	Imperial Valley-02, 5/19/1940, El Centro Array #9, 270
20-story	1	CHICHI/ILA013-W.at2	3.07	Chi-Chi Taiwan, 9/20/1999, ILA013, W
	2	IMPVALL/H-E07140.at2	2.31	Imperial Valley-06, 10/15/1979, El Centro Array #7, 140
	3	SUPERST/B-PTS315.at2	2.48	Superstition Hills-02, 11/24/1987, Parachute Test Site, 315
	4	WESTMORL/PTS315.at2	3.43	Westmorland, 4/26/1981, Parachute Test Site, 315
	5	CHICHI/TCU076-N.at2	2.23	Chi-Chi Taiwan, 9/20/1999, TCU076, N
	6	CHICHI/TCU071-E.at2	2.38	Chi-Chi Taiwan, 9/20/1999, TCU071, E
	7	DUZCE/BOL000.at2	2.39	Duzce Turkey, 11/12/1999, Bolu, 0
	8	DENALI/ps10047.at2	1.06	Denali Alaska, 11/3/2002, TAPS Pump Station #10, 47
	9	LOMAP/HCH090.at2	2.73	Loma Prieta, 10/18/1989, Hollister City Hall, 90
	10	CAPEMEND/CPM090.at2	3.47	Cape Mendocino, 4/25/1992, Cape Mendocino, 90
	11	ERZIKAN/ERZ-EW.at2	1.65	Erzican Turkey, 3/13/1992, Erzincan, EW

available. The total number of basic RVs for the three buildings are found to be $k =$ 49, 97, and 121, respectively. Using the sensitivity analysis, the reduced number significant of RVs k_R are considered to be 8 for the three buildings.

To verify REDSET, the 3-story frame is considered first. It is excited by the first ground motion listed in Table 6.13 (Chi-Chi Taiwan, 9/20/1999, TCU055, N) with SF 3.93. The reliability index β, the corresponding p_f, and TNDFEA required to implement the algorithm are summarized in Table 6.14 for both LSFs. As before, to study the simulation region, a parameter h is used. Denoting h^i and h^f are the values of h for the intermediate and the final iterations, two sets of β values are obtained and the information is summarized in Table 6.14. To evaluate the accuracy of the estimated p_f values, 600 000 cycles of MCS were conducted requiring about 1020 hours of continuous running of a computer. For comparison and verification, p_f values obtained by MCS are also shown in Table 6.14. The p_f values obtained by REDSET and MCS are very similar. However, it required only 199 TNDFEA when REDSET is used instead of 600 000 when MCS is used. For this example, by reducing the simulation region, the estimation of the reliability index improves slightly. As mentioned earlier, the simulation region may not be an important factor when an IRS is generated using KM. In any case, the results indicate that REDSET correctly predicted the p_f values and the algorithm is verified.

Table 6.14 Verification of REDSET using CHICHI/TCU055-N.at2 ground motion with 3.93 SF.

LSFs			REDSET	MCS
$h^i = 1$ $h^f = 1$	Inter-story drift	β	2.898	2.996
		p_f (E−3)	1.8768	1.3683
		TNDFEA	199	600 000
	Overall drift	β	3.312	3.275
		p_f (E−4)	4.6246	5.2833
		TNDFEA	182	600 000
$h^i = 1$ $h^f = 0.5$	Inter-story drift	β	2.975	2.996
		p_f (E−3)	1.4634	1.3683
		TNDFEA	199	600 000
	Overall drift	β	3.310	3.275
		p_f (E−4)	4.6665	5.2833
		TNDFEA	182	600 000

h^i and h^f are the values of the h parameter for intermediate and final iterations, respectively.

Table 6.15 Safety Index β for a suite of selected ground motions.

Record Number	Inter-story drift			Overall drift		
	3-story	9-story	20-story	3-story	9-story	20-story
1	2.975	2.104	4.302	3.310	2.596	4.941
2	3.802	3.379	5.985	4.068	3.688	6.175
3	3.464	2.474	4.149	3.763	2.850	4.373
4	3.136	1.800	2.413	3.445	1.906	4.513
5	3.076	4.344	2.743	3.662	4.624	3.921
6	3.409	2.159	4.961	4.055	3.904	6.020
7	2.488	1.451	3.354	3.444	1.535	4.854
8	2.709	2.409	3.799	3.470	3.031	3.878
9	3.611	2.084	3.821	3.951	3.238	4.449
10	3.794	2.708	3.801	4.471	3.446	4.355
11	2.443	3.734	3.576	3.165	3.985	5.287
$\beta_{average}$	3.173	2.604	3.900	3.710	3.164	4.797

After the successful verification with a relatively smaller problem, the reliability index β for all three buildings excited by the corresponding 11 selected earthquake acceleration time histories are calculated using REDSET with Sampling Scheme MS3-KM and the results are summarized in Table 6.15. As expected, β values are not identical for 11 time histories. This is expected since the frequency contents of all 11 design time histories cannot be the same. However, the range of β values for each building will help the designer enormously to decide whether any modification in the design is required or not. At least, the owner will have the opportunity to decide if the range is acceptable or not. The average of β value for 11 time histories for each building is also estimated and shown in the table. Although the three buildings were designed by expert professionals, the 20-story building appears to be the strongest for both the serviceability LSFs and the 9-story building appears to be the weakest. The results also indicate that the inter-story drift is more critical than the overall lateral displacement in all cases. This observation was observed by the authors in other studies and widely reported in the literature. The case study considered in Section 6.3 also indicated a similar phenomenon. Inter-story drift needs to be checked more carefully.

To study the estimated β values further, they are plotted in Figures 6.9, 6.10, and 6.11, respectively, for the three buildings. The spread in the estimated β values cannot be overlooked. The spreads for both LSFs are relatively smaller and more

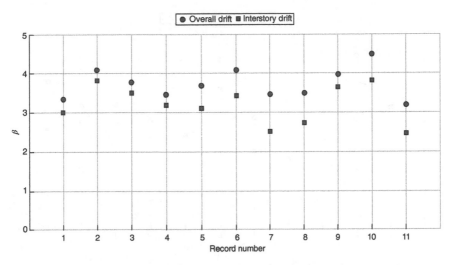

Figure 6.9 Safety index β for ground motions selected for the 3-story building for overall drift and inter-story drift.

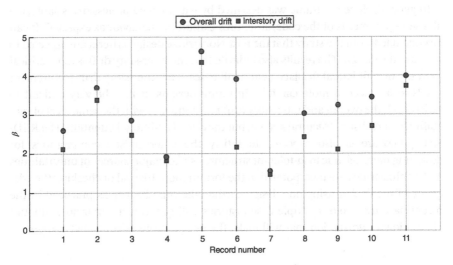

Figure 6.10 Safety index β for ground motions selected for the 9-story building for overall drift and inter-story drift

uniform for the 20-story building. For a structure with a higher period, this behavior is expected. These observations are expected due to the natural smoothness of the response spectrum for longer periods as indicated in Figure 4.5. For the selected ground motions, the plot of COVs appears to be smoother for longer periods.

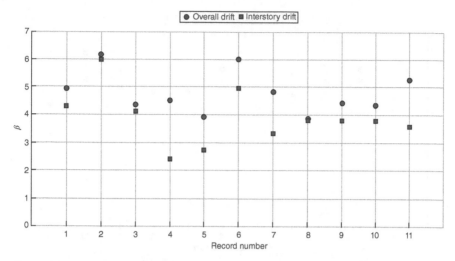

Figure 6.11 Safety index β for ground motions selected for the 20-story building for overall drift and inter-story drift.

In general, since the frame was designed by experienced professionals satisfying all the requirements of the code, structures exhibit the behavior as expected. It can be concluded from the study that the post-Northridge design criteria are expected to improve the design. The results also indicate that the inter-story drift is more critical than the overall lateral displacement in all cases. The inter-story drift requirement needs to be checked more carefully in future designs. It may be very difficult to satisfy both the overall and inter-story drift requirements with the same level of reliability for a realistic structural system, but they need additional attention. The RED-SET procedure is also expected to satisfy the deterministic communities for designing more seismic load-tolerant structures since major sources of uncertainties and nonlinearities are incorporated in the formulation. Instead of checking the adequacy of the design using one design earthquake time history as practiced in the recent past, the use of multiple time histories will generate an enormous amount of additional information that will make the design more seismic load-tolerant.

6.6 Concluding Remarks

The fundamental concept of risk assessment using REDSET was presented in Chapter 5. With the help of two examples, a case study and a set of three buildings designed by experts in the Los Angeles area, the REDSET concept is conclusively verified in this chapter for the earthquake loading. The robustness of the concept will be demonstrated by considering different types of dynamic loadings. Wave loading will be considered in Chapter 7 and thermomechanical loading will be considered in Chapter 9. Consideration of different computational platforms will be demonstrated in Chapter 8. A new design concept, known as PBSD, will be introduced in Chapter 8. The implementational potential of PBSD using SFEM representing structures with the assumed stress FE method discussed in Chapter 1 and REDSET will be demonstrated in this chapter.

All these different exercises are expected to established superior capabilities of REDSET. Some of the capabilities are not available to engineers at present to satisfy specified requirements in guidelines for designing complicated dynamic engineering systems. The evidences may convince the readers to use REDSET for the reliability estimation of CNDES excited in the time domain in the future.

7

Reliability Assessment of Jacket-Type Offshore Platforms Using REDSET for Wave and Seismic Loadings

7.1 Introductory Comments

REDSET was verified using examples and a case study in the previous chapter for onshore structures (ONS) excited by the earthquake loading. Multiple site-specific acceleration time histories were generated by fitting the TPGRS. It can be assumed at this time that REDSET was conclusively verified for the earthquake loading. It is now necessary to demonstrate the application potential of REDSET in estimating the risk of offshore structures (OFS), particularly the jacket-type offshore platforms (JTPs) excited separately by the wave and earthquake loadings applied in the time domain. Modeling of wave loading and the uncertainty associated with it were presented in Section 4.9. It was discussed that by using the constrained new wave (CNW) concept, an unlimited number of wave profiles could be generated in the time domain at a particular site. They contain the same meteoceanic properties, but their profiles and frequency contents are different.

Besides demonstrating the application potential of REDSET in estimating risk for OFS, specifically for JTPs, there are several other objectives for taking this example. JTPs can be considered a special class of structures. They need to be designed for both wave and earthquake loadings. In addition to the partial submergence-related issues, JPTs are constructed to satisfy special needs. Potential catastrophic environmental damages caused by the failure of a JTP can be enormous. However, the potential loss of human lives can be limited due to the advanced nature of the weather forecasting system currently available. No similar procedure for forecasting procedure is currently available for predicting the occurrence of a damaging earthquake in the near future. Research on related areas is still evolving. A comparative study of the behavior of a JTP excited separately by the wave and seismic loadings is expected to be extremely beneficial.

Since the dynamic characteristics of wave and earthquake loadings are very different, it will be very difficult to postulate which loading will be critical. Obviously,

Reliability Evaluation of Dynamic Systems Excited in Time Domain: Alternative to Random Vibration and Simulation, First Edition. Achintya Haldar, Hamoon Azizsoltani, J. Ramon Gaxiola-Camacho, Sayyed Mohsen Vazirizade, and Jungwon Huh.
© 2023 John Wiley & Sons, Inc. Published 2023 by John Wiley & Sons, Inc.

they need to be applied separately. In addition, earthquake excitation is applied at the base of the platform. However, the wave loading acts on the submerged elements and decays with the depth of the water from the free surface level. It may not be possible to make the underlying risks of failure identical for the two loadings, but attempts should be making them similar.

Another major objective of this example needs additional discussion. All the structures discussed in Chapter 6 are located in California; a very seismically active region in the United States. Information provided by several agencies in the United States was used to generate multiple acceleration time histories. Similar information may not be available for other parts of the world. The JTP considered in this example is located in the Persian Gulf region. Detailed information on the seismic activity of the region, similar to the United States, is not readily available. In fact, most of the world outside the United States is expected to be in this category. This example will demonstrate the seismic risk estimation procedure using REDSET in a region with limited information on seismic activity. A brief review of the modeling of wave loading discussed in Chapter 4 will be beneficial at this stage before moving forward.

7.2 Reliability Estimation of a Typical Jacket-Type Offshore Platform

A typical JTP, shown in Figure 7.1, was designed for the Persian Gulf region satisfying the American Petroleum Institute (API) standard (Dyanati and Huang 2014). The dimensions of the structure at the bottom are 29 m in the X direction and 36 m in the Y direction. The corresponding dimensions at the top are 14 m and 20 m, respectively. The frame is a 4-story structure with a total height of 76.25 m and the depth of the water is considered to be 65 m. The jacket legs are horizontally braced at four levels (-70.5, -48.25, -30.25, -12.5, and 5.75 m), as shown in Figure 7.1. Legs are made with hollow steel sections and the maximum and minimum diameters of 1.752 m and 1.651 m, respectively, and their corresponding areas are 0.3368 m^2 and 0.0669 m^2, respectively. The corresponding moments of inertia are 0.1202 m^4 and 0.0224 m^4, respectively. The total number of members in the structure is 237. The foundation for each leg of the platform is modeled using linear springs in the three orthogonal directions at the seabed level. The total mass acting at the top of the structure is considered to be 3.2×10^6 kg.

A finite element representation of the platform is shown in Figure 7.2 indicating members and joints by numbers. They are represented by beam-column elements. A member represented by Element 146 (Member 1 in Figure 7.1) is in the splash zone region. Another member represented by Element 95 (Member 2 in Figure 7.1) is located below the mean sea water level. Their reliabilities are estimated for the strength LSFs and will be discussed in more detail later.

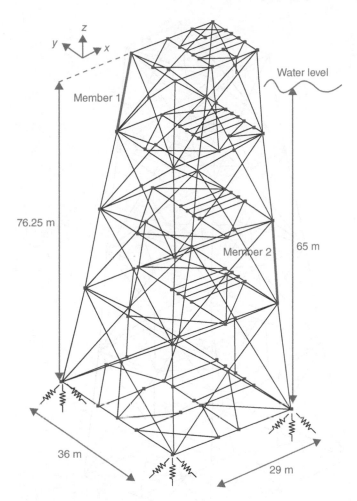

Figure 7.1 General sketch of the platform.

7.3 Uncertainty Quantifications of a Jacket-Type Offshore Platform

As discussed earlier, the statistical characteristics of all RVs present in the formulation must be collected before initiating any risk evaluation study. The quantifications of uncertainty in the resistance-related variables and wave loading are discussed separately in the following sections.

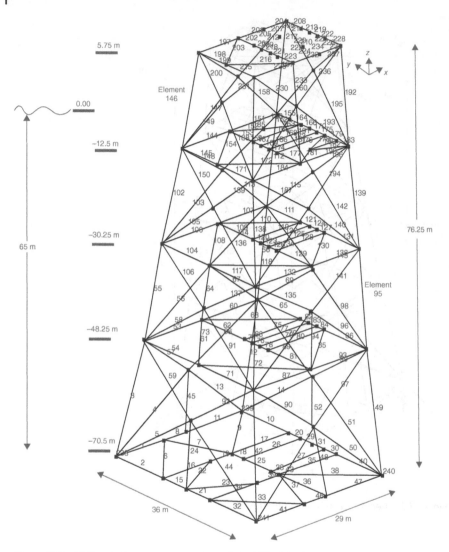

Figure 7.2 Finite element representation of the JTP with elements and joint numbers.

7.3.1 Uncertainty in Structures

As discussed earlier, ONS and OFS are very similar in terms of general structural arrangements. It is reasonable to assume that the uncertainty in the resistance-related variables in them would remain the same. Since the information on uncertainty on them for ONS is widely available, it can be used for OFS. However, uncertainty associated with some design variables peculiar to OFS, such as the

Table 7.1 Statistical information for resistance- and load-related RVs.

RV Classification	Random variable	Unit	Mean	COV	Distribution	References
Significant wave height	H_s	m	1.148	0.81	Weibull	Veritas (2010)
Hydrodynamic coefficients	C_M		1.2	0.1	Lognormal	API (2007)
	C_D		1.05	0.2	Lognormal	API (2007)
Marine growth	D_m	cm	7.5	0.5	Lognormal	Skallerud and Amdahl (2002)
Current velocity	V	m/s	0.8	0.2	Lognormal	Cassidy (1999)
Directionality	S		3	0.05	Uniform	ISO-19902 (2007)
Modulus of elasticity	E	GPa	206	0.03	Lognormal	JCSS (2006)
Yield stress of legs	f_y	MPa	335	0.07	Lognormal	JCSS (2006)
Strain-hardening ratio	b		0.02	0.1	Lognormal	Azizsoltani et al. (2018)

marine growth or hydrodynamic damping, the available information may be limited or experiments may need to be conducted or collected from past observations or investigations. Primary RVs in JTPs considered in this study include the modulus of elasticity, yield stress of steel, strain-hardening ratio b, areas, and moment of inertia of the members. Additional RVs that need to be considered for JTPs include the hydrodynamic coefficients C_M and C_D (drag and inertia coefficients), the marine growth D_m, the directionality, and the velocity of the current. The information on the design variables is summarized in Table 7.1. The reported mass at the top of the structures is considered to be the mean value. The distribution of it is considered to be lognormal with a COV of 0.1 (Ellingwood et al. 1980). The nominal areas and moments of inertia of the members are considered to be the mean values and the underlying distribution is lognormal with a COV of 0.05. The information on uncertainties is collected from the available literature. For ready reference, the sources are also cited in the table.

7.3.2 Uncertainty in Wave Loadings in the Time Domain

As discussed in detail in Chapter 4, many hours of random wave loading in the time domain can be simulated in a more computationally efficient manner using the CNW concept. In fact, Figure 4.17 is generated using the CNW theory. Using this procedure, five different random sea wave states for the site in the Persian Gulf region are generated. At present, consideration of multiple wave states is not required. However, to facilitate comparison with the seismic loading, five wave states are considered in this example.

7.4 Performance Functions

For the JTP shown in Figure 7.1, two types of LSF are considered: total lateral drift at the top of the platform and the strength of individual members.

7.4.1 LSF of Total Drift at the Top of the Platform

The LSF total drift is defined as:

$$g(\mathbf{X}) = \delta_{\text{allow}} - \hat{g}(\mathbf{X}) \tag{7.1}$$

where $\delta_{\text{allow}}$the allowable drift at the deck level and $\hat{g}(\mathbf{X})$ is the IRS for the drift at the deck level generated by REDSET using Sampling Scheme MS3-KM as discussed in Chapter 5. To generate the serviceability LSF, the information on the allowable deflection is necessary. This is an unexplored area for OFS. After a comprehensive literature review, the allowable deflection for JTPs suggested by Golafshani et al. (2011a) and Sharifian et al. (2015) and the following common practice for ONS steel structures (AISC 2011, 2017) are found to vary between $H/100$ and $H/600$, where H is the height of the structure. For this study, the maximum allowable deflection at the deck level is considered to be $H/200$ or 0.381 m.

7.4.2 Strength Performance Functions

The strength-related LSF considered is the failure of a member in strength. All the structural members shown in Figure 7.1 are considered to be made of steel in this study. Each member is subjected to both the axial load and bending moments. The interaction equations suggested by AISC (2011, 2017) for steel structures are considered for the strength of LSF. For the three-dimensional (3-D) JTPs, they can be represented as:

When $\dfrac{P_r}{P_c} \geq 0.2$

$$g(\mathbf{X}) = 1.0 - \left(\frac{P_r}{P_c} + \frac{8}{9}\frac{M_{rx}}{M_{cx}} + \frac{8}{9}\frac{M_{ry}}{M_{cy}}\right) = 1.0 - [\hat{g}_P(\mathbf{X}) + \hat{g}_M(\mathbf{X})]; \tag{7.2}$$

and

When $\dfrac{P_r}{P_c} < 0.2$

$$g(X) = 1.0 - \left(\dfrac{P_r}{2P_c} + \dfrac{M_{rx}}{M_{cx}} + \dfrac{M_{ry}}{M_{cy}}\right) = 1.0 - [\hat{g}_P(X) + \hat{g}_M(X)]; \qquad (7.3)$$

where P_r is the required axial strength, P_c is the available axial strength, M_{rx} and M_{ry} are the required flexural strength about the X and Y axes, respectively, and M_{cx} and M_{cy} are the available flexural strength of the X and Y axes, respectively. $\hat{g}_P(X)$ and $\hat{g}_M(X)$ are the IRSs for the axial load and the bending moment, respectively. The procedure is discussed in more detail in Section 3.13.3.

7.5 Reliability Evaluation of JTPs

Using REDSET with Sampling Scheme MS3-KM, the necessary IRSs and the corresponding serviceability and strength LSFs are developed. As indicated in Table 7.1, the significant wave height, H_s, is considered to be represented by the Weibull distribution as in Eq. 4.21. The two parameters of the distribution for the Persian Gulf region are $\alpha = 1.23$ and $\beta = 1.24$ (Veritas 2010).

Considering the JTP shown in Figure 7.1 and excited by the 3D wave loading, the total number of RVs present in the problem are found to be 72 or $k = 72$. After the sensitivity analysis, the most significant random variables are selected to be 5 or $k_R = 5$. REDSET with Sampling Scheme MS3-KM is used to estimate the reliability index β and the corresponding probability of failure p_f for both the serviceability and strength LSFs. The information is summarized in Table 7.2.

The total numbers of nonlinear dynamic finite element analyses (TNDFEA) required to implement REDSET are also given in the table. For the ready reference, TNDFEA is calculated as $(2k + 1) + (n - 2)(2k_R + 1) + 2^m + 2k_R + 1$, where $k = 72$, $k_R = 5$, $n = 4$, and $m = 4$, for both LSFs. The results were verified using 50 000 MCS cycles. High-performance computing (HPC) facilities available at the University of Arizona and parallel processing techniques were used to implement REDSET and to carry out the verification using MCS.

Based on the information given in Table 7.2, it can be observed that the results obtained by REDSET and MCS are similar. This observation clearly validates the REDSET concept. However, instead of using 50 000 simulation cycles, only about 200 TNDFEA are needed to extract the reliability information using REDSET. The reliability index β values estimated for both the total overall lateral drift and the strength LSFs are slightly different. For this JTP, β values for the overall serviceability LSF are slightly higher than the strength LSF indicating the platform is laterally stiff enough not to produce excessive overall lateral drift

Table 7.2 Structural reliability of OFS subjected to wave loading.

Wave states	LSF	Element No.	TDNFEA	REDSET with MS3-KM β	p_f	MCS β	p_f
1	Strength	146	182	2.755	0.00294	2.770	0.00280
		95	182	2.808	0.00249	2.794	0.00260
	Drift	[a]	194	3.180	0.00074	3.187	0.00072
2	Strength	146	182	2.385	0.00855	2.385	0.00854
		95	182	2.560	0.00523	2.528	0.00574
	Drift	[a]	194	2.956	0.00156	2.964	0.00152
3	Strength	146	182	2.703	0.00344	2.722	0.00324
		95	182	2.799	0.00256	2.785	0.00268
	Drift	[a]	194	2.897	0.00188	2.901	0.00186
4	Strength	146	182	2.384	0.00857	2.399	0.00822
		95	182	2.738	0.00310	2.718	0.00328
	Drift	[a]	194	2.982	0.00143	2.985	0.00142
5	Strength	146	182	2.485	0.00648	2.497	0.00626
		95	182	2.507	0.00609	2.474	0.00668
	Drift	[a]	194	2.840	0.00225	2.866	0.00208

[a] For total drift; lateral deflection at the top of the deck level.

or the serviceability LSF will not be critical for this example. Different β values for 5 wave states are also different as expected indicating the influence of frequency contents in them. However, the differences are not as significant as observed for the seismic loading (Gaxiola-Camacho 2017; Azizsoltani et al. 2018) and discussed in Chapter 6. This behavior is also expected considering the nature of irregularities in the wave and the earthquake loading. This will be discussed in more detail in Section 7.6.

The Box and Whisker plot of the data in Table 7.2 is shown in Figure 7.3. The plot shows the first quantile, third quantile, mean, median, minimum, and maximum of the data for β, and the corresponding LSF. The plots indicate that the p_f values for the overall drift are lower than the strength LSFs of Elements 146 and 95. The spread in the β values is the smallest for the overall drift and the largest for the strength LSF for the member in the splash zone (Element 146). It indicates that the modeling of waves in the splash zone is more challenging.

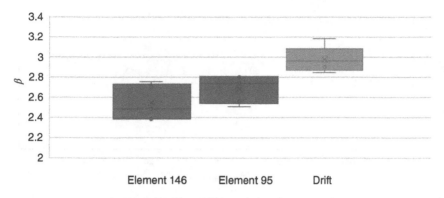

Figure 7.3 Summary of β values for different LFSs.

7.6 Risk Estimations of JTPs Excited by the Wave and Seismic Loadings – Comparison

One of the major advantages of reliability-based analysis and design is to compare the information on the risk or reliability of systems in different environments or for different design alternatives. This attractive feature can very efficiently be demonstrated using REDSET. The JTP shown in Figure 7.1 was designed for wave loading. It is now possible to estimate seismic risk of it. However, before checking the seismic risk, it is necessary to discuss the differences between designing structures in air and in the submerged condition. Even though both wave and earthquake loadings fall into the category of dynamic lateral loading, they act very differently. The wave force is applied to all members which are below the water surface level. This force is usually higher closer to the water surface and attenuated with the depth. It also needs to be pointed out that it may be difficult to figure out which loading will be more critical in a particular location and for a particular structure. The sea water level can be irregular similar to earthquake time histories. However, the frequency contents of the two loadings are expected to be very different. In this study, the seismic loading is modeled in two-dimension (2D). In the previous section, wave loading was applied in 3D. For the appropriate comparison, both loadings are considered in 2D in this section. Typical appearances of the wave and seismic loadings in the time domain in 2D are shown in Figures 7.4 and 7.5, respectively to demonstrate the nature of irregularities in them.

They look irregular. However, the time history for the seismic loading is much more irregular than wave loading. The frequency contents of the seismic loading are much higher than wave loading indicating that the dynamic responses of a structure exited in the time domain by the wave and seismic loadings are expected to be different. Furthermore, the dynamic properties (frequencies, mode shapes, etc.)

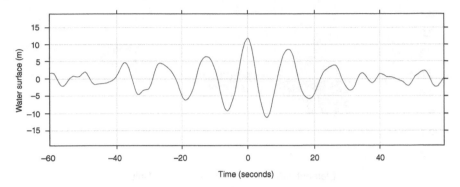

Figure 7.4 Sample water surface level using CNW – 2D representation.

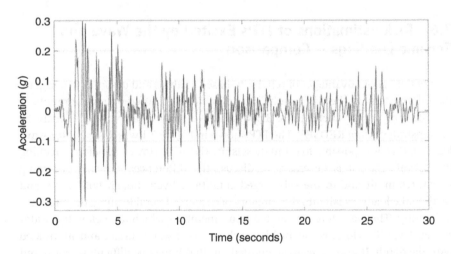

Figure 7.5 Time history of the 1940 Imperial Valley earthquake at El Centro station.

of the same structure vibrating in air and water are also expected to be very different. The natural frequencies of a JTP vibrating in water are expected to be lower than that of when it is vibrating in air. The responses of JTPs are expected to be significantly influenced by the presence of hydrodynamic damping (proportional to the velocity square) due to submergence instead of viscous damping (proportional to the velocity) commonly used for ONS vibrating in air.

As discussed earlier, the frequency contents of seismic and wave loadings are expected to be different. The dynamic properties of a structure in air and in water are also expected to be different. The lower frequency of wave loading and expected lower frequency of a structure vibrating in water caused by the submergence have the potential to cause very adverse responses as they approach the resonance condition. Moreover, the selection of design seismic and wave loadings can be very error-prone. Just before failure, the structure may develop many major sources of nonlinearity. The sources of nonlinearity and uncertainty need to be addressed explicitly. The related issues are very complicated and difficult, and often overlooked. They are comprehensively addressed here.

A casual literature review will indicate that OFS are being built in large numbers to address energy-related issues at present. There is no doubt that the failure of ONS can be enormous, particularly when the loss of human lives needs to be considered. If environmental damages are included, the failure of OFS can also be catastrophic in nature.

Since the structure is located in the Gulf region, the process of generating multiple design earthquake time histories discussed in Chapters 4 and 6 cannot be followed. Some of the readers would like to know how to generate design time histories in places where they reside. The necessary information required to generate them and available in the United States is not available for the site of interest. Several related issues are addressed in this section considering the application potential of REDSET for estimating seismic risk in all regions of the world. This section is expected to be very informative particularly when underlying risks of a structure vibrating in air and in water are compared.

As discussed in Chapter 4, the time histories of wave and earthquake loadings are very different. They also act very differently on a structure. The earthquake loading acts at the base of a structure and all masses are excited by the same amount of forces. Wave loading is estimated at the mean sea level, and it attenuates with depth. Structural elements at the splash zone are exposed to the most severe loading condition but are difficult to model. Since the basic structure considered in the previous section remains the same for this comparative study, no additional discussion is necessary to quantify uncertainty in all the RVs. Since earthquake loading is applied in two dimensions (2-D) in this section, wave loading also needs to be similarly applied in 2-D to make the comparison more realistic. Also, since it is virtually impossible that the critical wave and earthquake loadings will act simultaneously, they need not be applied simultaneously for the design of OFS.

The current seismic design in the United States requires considering at least 7 or 11 time histories. Although it is not explicitly stated in a design guideline for OFS, five different wave profiles in the time domain are considered to satisfy the intent of using multiple time histories to consider a significant amount of uncertainty in modeling them, similar to the seismic design guidelines.

Since the JTP considered here is located in the Persian Gulf region, the generation of the site-specific TPGRS according to ASCE (2010), discussed in Chapter 4, is not possible. It is also difficult to obtain design details of a JTP that was also designed for earthquake loading. The procedure suggested by the International

Standard Organization (ISO) is used to generate multiple earthquake time histories for the Persian Gulf region, as briefly discussed next. This exercise may assist the readers to generate multiple earthquake acceleration time histories outside the United States with limited information on seismic activity.

ISO-19901-2 (2017) suggests two levels of earthquakes: Extreme-level earthquake (ELE) and abnormal-level earthquake (ALE). ALE is of interest in this study. For this level of excitation, the structures are expected to develop nonlinear behavior and some damages are expected; however, they should not suffer a complete loss of integrity or collapse. According to ISO-19901-2 (2017), the Gulf region is in Seismic Zone 2 with the spectral acceleration at the period of 1 and 0.2 seconds are $S_a(T = 1) = 0.3$ g and $S_a(T = 1) = 0.75$ g, respectively.

According to ISO-19900 (2013), manned OFS should be designed for the life-safety category of S1 for seismic loading. They should also be considered in the consequence category of C2 with the corresponding exposure level of L1. For Seismic Zone 2 and exposure level L1, the Seismic Risk Category (SRC) is 4. Assuming that the piles at the supports of the JTP will produce the horizontal ground motions near the surface soil (API 2007), site class is assumed to be D. Based on site class as well as $S_a(T = 1)$ and $S_a(T = 0.2)$, the site coefficients for the acceleration and velocity of the response spectrum, denoted as C_a and Cv, respectively, are estimated to be 1 and 1.2. The ISO 1000-year horizontal acceleration spectrum at the site $S_{a,\text{site}}(T)$ can be expressed as:

$$S_{a,\text{site}}(T) = \begin{cases} (3T + 0.4)C_a S_{a,\text{map}}(0.2); \ T \le 0.2\,\text{s} \\ \min\left[C_a S_{a,\text{map}}(0.2), C_v S_{a,\text{map}}(1.0)/T \right]; 0.2\,\text{s} < T \le 4\,\text{s} \\ 4C_v S_{a,\text{map}}(1.0)/T; \ T > 4\,\text{s} \end{cases} \quad (7.4)$$

where T is the fundamental period of the structure. The acceleration response spectrum for ALE earthquake, $S_{a,\text{ALE}}(T)$, can be obtained by taking the product of a scale factor based on the structure exposure level, N_{ALE} and $S_{a,\text{site}}(T)$, as:

$$S_{a,\text{ALE}}(T) = N_{\text{ALE}} \times S_{a,\text{site}}(T) \quad (7.5)$$

Considering the exposure level L1, N_{ALE} is estimated to be 1.6. The ELE spectral acceleration can also be calculated as:

$$S_{a,\text{ELE}}(T) = S_{a,\text{ALE}}(T)/C_r \quad (7.6)$$

where C_r is seismic reserved capacity factor. For steel JTPs based on the characteristics considered in this example, C_r is estimated to be 2.8. The site-specific spectra for the location of the structure, ALE and ELE are shown in Figure 7.6.

Obviously, any number of time histories can be used to fit a spectrum. As mentioned earlier, at least 11 acceleration time histories are required to be considered as discussed in Chapter 6. As discussed in Chapter 4, the entire PEER database

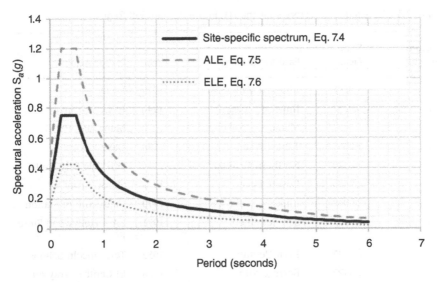

Figure 7.6 Site-specific ALE and ELE spectra.

consisting of several thousands of earthquake time histories recorded from all over the world is used to select a suite of ground motions, denoted as Alternative 1 in Section 4.3. Following the ground motion selection process discussed earlier, the database can be reduced from several thousand to several hundred.

Considering ALE as the target spectrum, at least 11 scaled earthquake time history records need to be selected. The submerged fundamental period of the structure is estimated as 2.1 seconds. The intention is to scale the records so that their spectral values match the ALE spectrum at a fundamental period of the structure of 2.1 seconds.

A suitability factor concept discussed in detail in Section 4.3 is used to select the most suitable site-specific 11 time histories. The procedure is not repeated here. The selected earthquake time histories for the JTP located in the Persian Gulf region are listed in Table 7.3. They are expected to satisfy the intent of ASCE/SEI (2010).

The response spectra for all selected 11 earthquakes are shown in Figure 7.7. They matched exactly at the fundamental period of 2.1 seconds of the JTP. The target response spectrum is shown in a solid black line. The mean value and the ±1 standard deviation range of these spectra are also shown in dotted lines in the figure. The 11 selected earthquake time histories are expected to satisfy all the requirements for earthquake loading.

To make the comparison between the wave and seismic loadings more reasonable, the statistical characteristics of all the variables identified in Table 7.1

Table 7.3 Selected time histories for the JTP in the Persian Gulf Region.

Record number	Scale factor	Earthquake name	Year	Station name
1	1.0696	Imperial Valley-06	1979	Delta
2	0.9763	Imperial Valley-02	1940	El Centro Array #9
3	1.3497	Managua_ Nicaragua-01	1972	Managua_ ESSO
4	1.6083	Tabas_ Iran	1978	Dayhook
5	1.0292	Northern Calif-03	1954	Ferndale City Hall
6	2.9505	Coyote Lake	1979	Gilroy Array #4
7	2.8942	Imperial Valley-06	1979	Calexico Fire Station
8	2.0342	San Fernando	1971	LA – Hollywood Store FF
9	2.9332	Kern County	1952	Taft Lincoln School
10	1.3927	Borrego Mtn	1968	El Centro Array #9
11	3.4952	Friuli_Italy-02	1976	Buia

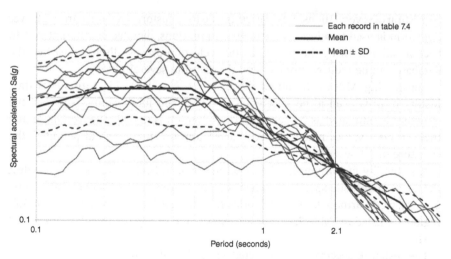

Figure 7.7 Selected ground motions, their mean values, one standard deviation minus and plus, and the target spectra.

for wave loading are considered to be the same for earthquake loading also. The velocity of the current is considered to be zero. Since the JTP is excited by 2D earthquake loading, wave loading is also applied in 2D for comparison.

Considering all the assumptions mentioned here, the total number of RVs are found to be 70 or $k = 70$. After the first iteration, the reduced number of RVs is

Table 7.4 Reliability estimation for wave loading.

Wave states	LSF	Element no.	REDSET with MS3-KM			MCS	
			TDNFEA	β	p_f	β	p_f
1	Strength	146	180	2.425	0.00764	2.429	0.00758
		95	180	2.712	0.00335	2.699	0.00348
	Drift	a	192	2.893	0.00191	2.915	0.00178
2	Strength	146	180	2.694	0.00353	2.695	0.00352
		95	180	2.705	0.00342	2.706	0.00340
	Drift	a	192	2.950	0.00159	2.964	0.00152
3	Strength	146	180	2.748	0.00299	2.754	0.00294
		95	180	2.743	0.00304	2.739	0.00308
	Drift	a	192	3.009	0.00131	3.011	0.00130
4	Strength	146	180	2.644	0.00410	2.664	0.00386
		95	180	2.643	0.00411	2.639	0.00416
	Drift	a	192	3.082	0.00103	3.090	0.00100
5	Strength	146	180	2.491	0.00638	2.495	0.00630
		95	180	2.567	0.00513	2.559	0.00540
	Drift	a	192	3.079	0.00104	3.079	0.00104

[a] For total drift; lateral deflection at the top of the deck level.

found to be 5 or $k_R = 5$. For the serviceability LSF, the allowable deflection at the top of the platform remained the same as 0.38125 m. The reliability index β and the corresponding probability of failure, p_f, for the strength and serviceability LSFs are estimated using REDSET with Sampling Scheme MS3-KM. The information on β, p_f, and the necessary TDNFEA for 5 wave states is summarized in Table 7.4 for the overall drift at the deck level and for the strength LSFs for Elements 146 and 95. The results are verified using 10 000 MCS using the HPC at the University of Arizona and the parallel processing techniques.

The reliability estimation results for the JTP excited by the wave loading summarized in Tables 7.2 and 7.4 are slightly different, as expected. Results shown in Tables 7.2 and 7.4 are for wave loading applied in 3D and 2D, respectively, as discussed previously in this section.

The same information on β and the corresponding p_f values for the JTP when excited by the selected 11 design earthquake time histories is summarized in Table 7.5 for the overall lateral deck deflection. The same information for the strength LSF for Elements 146 and 95 is summarized in Table 7.6. It is to be noted that Element 146 is in the splash zone.

Table 7.5 Reliability estimation for the overall drift at the deck level for the earthquake excitations.

Record number	REDSET with MS3-KM			MCS	
	TDNFEA	β	p_f	β	p_f
1	178	2.567	5.12E−03	2.576	5.00E−03
2	178	2.394	8.32E−03	2.366	9.00E−03
3	178	2.503	6.16E−03	2.506	6.10E−03
4	178	2.699	3.48E−03	2.727	3.20E−03
5	178	2.695	3.51E−03	2.727	3.20E−03
6	178	2.642	4.12E−03	2.652	4.00E−03
7	178	2.709	3.38E−03	2.697	3.50E−03
8	178	2.452	7.09E−03	2.437	7.40E−03
9	178	2.323	1.01E−02	2.297	1.08E−02
10	189	2.202	1.38E−02	2.212	1.35E−02
11	178	2.366	9.00E−03	2.342	9.60E−03

Similar to Figure 7.3, the Box and Whisker plots of the data in Tables 7.5 and 7.6 are shown in Figure 7.8 for wave loading and in Figure 7.9 for earthquake loading. The plots indicate that the p_f values for the overall drift are lower than the strength LSFs of Elements 146 and 95. The spread in the β values is the smallest for the overall drift and the largest for the strength LSF for Element 146 in the splash zone. It indicates that the modeling of waves in the splash zone is more challenging. The plots also indicate that there is more uncertainty in earthquake loading compared to wave loading. The strength failure is more likely than the failure caused by the excessive lateral drift for the JTP considered in the study.

7.7 Comparison of Results for the Wave and Earthquake Loadings

For both the wave and earthquake excitations, β or p_f values obtained by REDSET and MCS are very similar. However, REDSET requires only about 200 TDNFEA or deterministic FE analyses instead of 10 000 MCS for both the serviceability and strength LSFs. REDSET appears to be very efficient and accurate in estimating risk for both the seismic and wave loadings, confirming the observations made for the earthquake loading discussed in Chapter 6.

Table 7.6 Reliability estimation for strength LSF for Elements 146 and 95 for the earthquake excitations.

Record number	Element 146 (member 1) in Splash Zone					Element 95 (member 2)				
	REDSET with MS3-KM			MCS		REDSET with MS3-KM			MCS	
	TDNFEA	β	p_f	β	p_f	TDNFEA	β	p_f	β	p_f
1	189	2.503	0.00617	2.536	0.00560	189	2.410	0.00797	2.409	0.00800
2	189	2.164	0.0152	2.165	0.0152	189	2.277	0.0114	2.270	0.0116
3	189	2.306	0.0105	2.319	0.0102	189	2.392	0.00837	2.382	0.00860
4	189	2.443	0.00728	2.442	0.00730	189	2.543	0.00549	2.562	0.00520
5	178	2.854	0.00216	2.848	0.00220	178	2.373	0.00882	2.338	0.00970
6	189	2.338	0.00970	2.378	0.00870	189	2.391	0.00841	2.357	0.00920
7	189	2.860	0.00212	2.834	0.00230	189	2.404	0.00810	2.423	0.00770
8	189	2.472	0.00673	2.442	0.00730	189	2.236	0.0127	2.245	0.0124
9	189	2.299	0.0108	2.297	0.0108	189	2.042	0.0206	2.042	0.0206
10	189	2.311	0.0104	2.330	0.00990	189	2.150	0.0158	2.139	0.0162
11	189	2.272	0.0116	2.280	0.0113	189	2.359	0.00915	2.362	0.00910

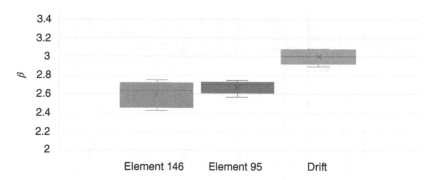

Figure 7.8 Summary of β values for different LFSs for 2D wave loading.

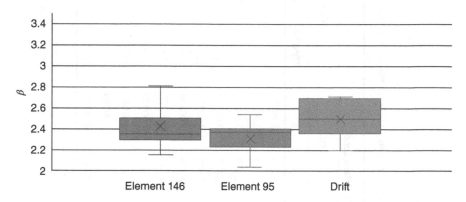

Figure 7.9 Summary of β values for different LFSs for earthquake loading.

Several other important observations can be made from the results obtained in this study. For this structure, the mean values of p_f for both the earthquake and wave loadings are of the same order of magnitude indicating that the JTP was designed reasonably well. For this particular JTP, the mean p_f values for serviceability and strength LSFs due to earthquake loading appear slightly higher than that of wave loading, indicating earthquake loading will control the design. In terms of risk and consequences of failure, earthquake loading is expected to be more crucial than wave loading since the occurrence of an earthquake is not predictable while it is possible to evacuate the platform in response to the approaching adverse weather conditions. Therefore, the structure should have higher safety factors against seismic loading. For the Gulf region where the JPT is located, several major faults are present and designing for more critical earthquake loading is reasonable.

Element 146 is in the splash zone. From Figure 7.4, it is observed that predicting its p_f in strength can be difficult as expected. COVs of p_f are observed to be higher for seismic loading than wave loading for both the strength and serviceability LSFs. This may indicate that there is more uncertainty in predicting seismic loading than wave loading. The period of wave loading is expected to be higher than earthquake loading. However, the submerged state of the OFS is expected to increase its period compared to when they are not submerged. This tendency of the wave period approaching the resonance condition may cause wave loading more critical than earthquake loading in some cases. However, it is not the case for the structure considered in the case study. It is very encouraging to note that REDSET can be used to estimate the risk of JTPs accurately and efficiently for different LSFs. Since no such method is currently available, the proposed method can be used for the uncertainty management of OFS and other CNDES.

7.8 Concluding Remarks

It is very encouraging to note that REDSET is capable of extracting underlying risk for both the earthquake and wave loadings for OFS. The comparison identified several important issues for designing ONS and OFS. Modeling of wave loading applied in the time domain is not routinely practiced in the profession. However, it is one of the major sources of uncertainty in designing OFS. To accurately model the uncertainty in wave loading, the fluctuation of water surface level was modeled in the time domain. The directionality of wave loading and its three-directional behavior were also considered. The study established that REDSET is robust and efficient in estimating the risk of CNDES when excited by the dynamic loadings in the time domain with very different characteristics. Generating multiple time histories for the design earthquake were generated in Chapters 4 and 6 by scaling recorded time histories available at the PEER database or by using BBP when the information on the seismic activities of the region of interest is readily available. Another alternative is proposed when available information is limited or when the procedure suggested in the US guidelines cannot directly be used. REDSET with Sampling Scheme MS3-KM appears to be applicable for all cases considered here. Based on the promising results of Chapters 6 and 7, it can be concluded that REDSET can be used for the reliability analyses of OFS and ONS exciting them in the time domain with wave and earthquake loadings.

8

Reliability Assessment of Engineering Systems Using REDSET for Seismic Excitations and Implementation of PBSD

8.1 Introductory Comments

Reliability estimation of CNDES using REDSET was introduced in Chapter 5 and conclusively verified in Chapters 6 and 7 for earthquake and wave loadings applied in the time domain. In Chapter 1, it was stated that the REDSET algorithm could be used in any computational platform to estimate the underlying risk of CNDES and it would replace the assumed stress-based SFEM concept developed over three decades ago by Haldar and Mahadevan (2000b). It was also mentioned that users could use any computer program available to them capable of conducting nonlinear dynamic analysis of structures excited in the time domain. Since it is very commonly used, structures were represented by the displacement-based finite element method (FEM) in all examples in the previous chapters. To emphasize the point that structures can be represented by any type of FEM, structures are represented by the assumed stress-based FEM in all the examples given in this chapter.

Since it is a relatively long chapter, some introductory comments are necessary to describe its contents. In the first part of this chapter, the stress-based FEM formulation for solving CNDSE excited in the time domain is briefly discussed. After the Northridge earthquake of 1994, flexible post-Northridge connections for steel frame structures were proposed. REDSET is used to document the improvement in the underlying risk of steel frames in the presence of flexible connections. To avoid considerable economic losses caused by the Northridge earthquake, a new design concept known as the *performance-based seismic design* (PBSD) was introduced. It replaces the older life safety (LS) concept so that designers/owners can design a structure for a specific performance level with quantifiable acceptable risk. However, PBSD did not specify how to quantify the underlying risk for specific performance levels. In fact, there is no method currently available for the estimation of the risk to implement the PBSD concept. It will be demonstrated with the help of examples that REDSET will fill this knowledge gap. Implementation of PBSD

Reliability Evaluation of Dynamic Systems Excited in Time Domain: Alternative to Random Vibration and Simulation, First Edition. Achintya Haldar, Hamoon Azizsoltani, J. Ramon Gaxiola-Camacho, Sayyed Mohsen Vazirizade, and Jungwon Huh.

using REDSET is showcased in this chapter to help practicing engineers use it with confidence. Historical evidences also suggest that in addition to the severity of an earthquake, soil condition at a site is very important in studying the vulnerability of structures. Structures built on poor soil suffer more damage. Thus, soil conditions at a site need to be correlated with the expected performance level for a structure. If an earthquake time history is selected for a specific performance level and soil condition, and the corresponding allowable value is also selected using PBSD criteria, the underlying risk should be similar for different performance levels and soil conditions. This may appear to be complicated but can be demonstrated with the help of REDSET.

All these major objectives are considered one by one in the following sections. To avoid any confusion, it is important to note that PBSD can be implemented by representing structures using either the displacement- or the assumed stress-based FEM and the underlying risk can be estimated by REDSET.

8.2 Assumed Stress-Based Finite Element Method for Nonlinear Dynamic Problems

Since this chapter is specifically developed using the assumed stress-based FEM, the basic concept for the nonlinear dynamic analysis of framed structures excited in the time domain is briefly discussed first followed by the solution strategies needed to execute it.

8.2.1 Nonlinear Deterministic Seismic Analysis of Structures

The basic review of the assumed stress-based FEM concept is presented below so that an IRS can be generated by using it and then REDSET can be used to extract reliability information. It will be established that any form of FEM representation of structures (displacement or assumed stress-based) to estimate nonlinear responses required to generate an IRS and the corresponding LSF integrated with REDSET will estimate the underlying risk very accurately and efficiently. The process will be very robust and can be executed on any computational platform. Steel structures are emphasized in the following discussions since PBSD guidelines are primarily developed for them at present.

8.2.2 Seismic Analysis of Steel Structures

Steel structures excited by the earthquake acceleration time histories develop nonlinear behavior at the very early stage of shakings. Several works related to nonlinear seismic analysis of steel structures can be found in the literature. It is not practical to cite most of them here but they are listed in the extensive References section of the book. The authors and their team members also published extensively on the efficiency, accuracy, and robustness of the assumed stress-based FEM in the presence of several sources of nonlinearities (geometric, material, and joint conditions) satisfying the underlying physics. The basic concept is briefly reviewed later.

8.2.3 Dynamic Governing Equation and Solution Strategy

Without losing any generality, for large structural systems, the nonlinear dynamic governing equation of motion in matrix notation is expressed as:

$$\mathbf{M}\ddot{\mathbf{D}}^{(n)}_{t+\Delta t} + \mathbf{C}_t\dot{\mathbf{D}}^{(n)}_{t+\Delta t} + \mathbf{K}^{(n)}_t \Delta \mathbf{D}^{(n)}_{t+\Delta t} = \mathbf{F}^{(n)}_{t+\Delta t} - \mathbf{R}^{(n-1)}_{t+\Delta t} - \mathbf{M}\ddot{\mathbf{D}}^{(n)}_{g,(t+\Delta t)} \qquad (8.1)$$

where $\mathbf{M}$, $\mathbf{C}_t$, and $\mathbf{K}^{(n)}_t$ represent the mass, damping, and global stiffness matrix of the nth iteration at time t, respectively; the vectors $\ddot{\mathbf{D}}^{(n)}_{t+\Delta t}$, $\dot{\mathbf{D}}^{(n)}_{t+\Delta t}$, $\Delta \mathbf{D}^{(n)}_{t+\Delta t}$, and $\ddot{\mathbf{D}}^{(n)}_{g,(t+\Delta t)}$ are the acceleration, velocity, incremental displacement, and ground acceleration of the nth iteration at time $t+\Delta t$, respectively; $\mathbf{F}^{(n)}_{t+\Delta t}$ is the external load vector of the nth iteration at time $t+\Delta t$; and $\mathbf{R}^{(n-1)}_{t+\Delta t}$ is the internal force vector of the $(n-1)$th iteration at time $t+\Delta t$.

To calculate every matrix and vector contained in Eq. 8.1, the following procedures can be used. The mass matrix can be represented as lumped or consistent. Since most of the structural elements in a frame-type structure are expected to carry both the axial load and bending moment, they are generally represented by beam-column elements. The tangent stiffness matrix and the internal force vector for every beam-column element can be expressed explicitly at a given time t and nth iteration considering both the geometric and material nonlinearities (Kondoh and Atluri 1987; Shi and Atluri 1988; Haldar and Mahadevan 2000b).

Since the mass and stiffness matrixes are readily available from the FE formulation, the Rayleigh-type damping will be appropriate and is used. The damping matrix can be expressed as (Leger and Dussault 1992):

$$\mathbf{C} = \alpha\,\mathbf{M} + \gamma\,\mathbf{K} \qquad (8.2)$$

where α and γ are the mass and stiffness proportional constants, respectively, and can be estimated from the first two natural frequencies of the structure (Clough and Penzien 1993; Lee and Haldar 2003b).

The Newmark's step-by-step direct integration numerical analysis procedure was used to solve Eq. 8.1. The following solution strategy was used. At every time step Δt, the displacement and velocity vectors can be expressed as (Bathe 1982):

$$\mathbf{D}_{t+\Delta t}^{(n)} = \mathbf{D}_t + \dot{\mathbf{D}}_t \Delta t + \left[\left(\frac{1}{2} - \beta_N\right)\ddot{\mathbf{D}}_t + \beta_N \ddot{\mathbf{D}}_{t+\Delta t}^{(n)}\right]\Delta t^2 \tag{8.3}$$

and

$$\dot{\mathbf{D}}_{t+\Delta t}^{(n)} = \dot{\mathbf{D}}_t + \left[(1-\eta)\ddot{\mathbf{D}}_t + \eta\ddot{\mathbf{D}}_{t+\Delta t}^{(n)}\right]\Delta t \tag{8.4}$$

where β_N and η are parameters required in the Newmark Beta method to obtain integration stability and accuracy. In this study, $\beta_N = 1/4$ and $\eta = 1/2$ were used. For these values at every Δt interval, the acceleration vector is constant, and the procedure is expected to be stable.

Hence, the acceleration at time $t + \Delta t$ and nth iteration can be expressed as:

$$\ddot{\mathbf{D}}_{t+\Delta t}^{(n)} = \frac{1}{\beta_N}\left[\frac{1}{\Delta t^2}\left(\mathbf{D}_{t+\Delta t}^{(n)} - \mathbf{D}_t - \dot{\mathbf{D}}_t \Delta t\right) - \left(\frac{1}{2} - \beta_N\right)\ddot{\mathbf{D}}_t\right] \tag{8.5}$$

And the velocity at time $t+\Delta t$ and nth iteration can be derived as:

$$\dot{\mathbf{D}}_{t+\Delta t}^{(n)} = \dot{\mathbf{D}}_t + \left[\left(1 - \frac{\eta}{2\beta_N}\right)\ddot{\mathbf{D}}_t + \frac{\eta}{\beta_N \Delta t^2}\left(\mathbf{D}_{t+\Delta t}^{(n)} - \mathbf{D}_t - \dot{\mathbf{D}}_t \Delta t\right)\right]\Delta t \tag{8.6}$$

The displacement and dynamic force vectors at time $t+\Delta t$ and nth iteration can be expressed in incremental form as:

$$\mathbf{D}_{t+\Delta t}^{(n)} = \mathbf{D}_{t+\Delta t}^{(n-1)} + \Delta\mathbf{D}_{t+\Delta t}^{(n)} \tag{8.7}$$

and

$$\mathbf{F}_{t+\Delta t}^{(n)} = \mathbf{F}_{t+\Delta t}^{(n-1)} + \Delta\mathbf{F}_{t+\Delta t}^{(n)} \tag{8.8}$$

Substituting Eqs. 8.2, 8.5, and 8.6 into Eq. 8.1, manipulating by assembling similar terms together, and using Eqs. 8.7 and 8.8, the nonlinear dynamic governing equation can be expressed as:

$$\mathbf{K}_D^t \Delta\mathbf{D}_{t+\Delta t}^{(n)} = \mathbf{F}_{D(t+\Delta t)}^{(n)} - \mathbf{R}_{t+\Delta t}^{(n-1)} \tag{8.9}$$

where $\mathbf{K}_D^t$ is the dynamic tangent stiffness matrix of the system at time t, $\Delta\mathbf{D}_{t+\Delta t}^{(n)}$ and $\mathbf{F}_{D(t+\Delta t)}^{(n)}$ are the incremental displacement vector and modified external load vector of the nth iteration at time $t + \Delta t$, respectively, and $\mathbf{R}_{t+\Delta t}^{(n-1)}$ was defined earlier.

The dynamic tangent stiffness matrix can be expressed as:

$$\mathbf{K}_D^t = f_1\mathbf{M} + f_2\mathbf{K}^t \tag{8.10}$$

And the modified external load vector $\mathbf{F}_{D(t + \Delta t)}^{(n)}$ can be expressed as (Lee and Haldar 2003a):

$$\mathbf{F}_{D(t + \Delta t)}^{(n)} = \mathbf{F}_{D(t + \Delta t)}^{(n-1)} + \Delta \mathbf{F}_{D(t + \Delta t)}^{(n)} \tag{8.11}$$

where $\mathbf{F}_{D(t + \Delta t)}^{(n-1)}$ is the modified external force vector of the $(n - 1)$th iteration at time $t + \Delta t$; and $\Delta \mathbf{F}_{D(t + \Delta t)}^{(n)}$ is the incremental modified external force vector of the nth iteration at time $t + \Delta t$.

The incremental modified external force vector $\Delta \mathbf{F}_{D(t + \Delta t)}^{(n)}$ of the nth iteration at time $t + \Delta t$ can be represented as:

$$\Delta \mathbf{F}_{D(t + \Delta t)}^{(n)} = \Delta \mathbf{F}_{t + \Delta t}^{(n)} - \mathbf{M} \Delta \ddot{\mathbf{D}}_{g,(t + \Delta t)}^{(n)} \tag{8.12}$$

And the modified external vector $\mathbf{F}_{D(t + \Delta t)}^{(n-1)}$ of the $(n - 1)$th iteration at time $t + \Delta t$ can be expressed as:

$$\mathbf{F}_{D(t + \Delta t)}^{(n-1)} = \mathbf{F}_{t + \Delta t}^{(n-1)} + \mathbf{P}_{t + \Delta t}^{(n-1)} - \mathbf{M} \ddot{\mathbf{D}}_{g,(t + \Delta t)}^{(n-1)} \tag{8.13}$$

where $\mathbf{P}_{t + \Delta t}^{(n-1)}$ is the modified force vector contributed by the displacement, velocity, and acceleration at time t and the displacement vector at time $t + \Delta t$. It can be represented as:

$$\mathbf{P}_{t + \Delta t}^{(n-1)} = \mathbf{M}\left[f_1 \mathbf{D}_t + f_3 \dot{\mathbf{D}}_t + f_4 \ddot{\mathbf{D}}_t - f_1 \mathbf{D}_{t + \Delta t}^{(n-1)} \right]$$
$$+ \mathbf{K}^t \left[f_5 \mathbf{D}_t + f_6 \dot{\mathbf{D}}_t + f_7 \ddot{\mathbf{D}}_t - f_5 \mathbf{D}_{t + \Delta t}^{(n-1)} \right] \tag{8.14}$$

where the coefficients f_is are constants evaluated in terms of α, γ, η, β_N, and Δt as (Lee and Haldar 2003a):

$$f_1 = \frac{1}{\beta_N \Delta t^2} + \frac{\eta \alpha}{\beta_N \Delta t} \tag{8.15}$$

$$f_2 = \frac{\eta \gamma}{\beta_N \Delta t} + 1 \tag{8.16}$$

$$f_3 = \frac{1}{\beta_N \Delta t} + \frac{\eta \alpha}{\beta_N} - \alpha \tag{8.17}$$

$$f_4 = \left(\frac{1}{2\beta_N} - 1 \right) + \eta \alpha \left(\frac{1}{2\beta_N} - \frac{1}{\eta} \right) \Delta t \tag{8.18}$$

$$f_5 = \frac{\eta \alpha}{\beta_N \Delta t} \tag{8.19}$$

$$f_6 = \frac{\eta\gamma}{\beta_N} - \gamma \tag{8.20}$$

$$f_7 = \left(\frac{\eta\gamma}{2\beta_N} - \gamma\right)\Delta t \tag{8.21}$$

Equation 8.9 was solved using the modified Newton-Raphson method with arc-length control (Lee and Haldar 2003a). Using this FE algorithm, the deterministic responses of structures excited by the seismic loading applied in the time domain can be estimated very efficiently and accurately for frame-type steel structures. A very sophisticated computer program was developed and verified. It is still used to verify dynamic responses obtained by other computer programs currently used. Any detailed information on the topic can be found in Haldar and Mahadevan (2000b).

8.2.4 Flexibility of Beam-to-Column Connection Models by Satisfying Underlying Physics – Partially Restrained Connections for Steel Structures

The earlier formulation is essentially for all beam-to-column connections to be fully restrained (FR) type. In the United States, three basic types of connections are used, specifically for steel frames. Type I is commonly used in rigid frames. It is known as FR type and assumes that beam-to-column connections have sufficient rigidity to hold the original angles between intersecting members virtually unchanged. Type II is used in simple frames and assumes that the ends of the beams and girders are connected by shear only and free to rotate under the gravity load. Type III is commonly known as the semirigid or partially restrained (PR) type. It assumes that the connections possess a quantifiable and known moment capacity intermediate in degrees between the rigidity of Type I and the flexibility of Type III. Experimental studies suggest that all connections for steel structures are flexible with different degrees of rigidity. Thus, any algorithm used for the nonlinear dynamic analysis needs to be modified to consider the presence of PR connections.

The behavior of a PR connection is generally described in terms of the relationship between the moment transmitted by the connection, M, and the relative rotation angle, θ, of connecting members. This relationship is commonly represented by M–θ curves. There are many analytical models to represent PR connections. In the past, models representing M–θ relationships of connections were assumed to be linear. Generally, linear models depend on the initial stiffness, which represents the connection behavior for the entire range of loading. However, since the stiffness of the connection decreases as the moment increases, linear models represent a crude approximation of real M–θ relationships of connections. To address this issue, bilinear models were introduced (Lionberger and Weaver 1969). They combined initial stiffness with a lower stiffness at a specific transition moment. Unfortunately, linear and bilinear models provide a rudimentary representation of M–θ relationships of connections. As an alternative, the piecewise linear model can be used. This representation is very similar to the bilinear model; the only difference is that the piecewise linear model uses several straight lines to represent the M–θ relationship. Linear, bilinear, and piecewise linear models are relatively easy to use. However, discontinuity in connection stiffness may cause numerical problems, particularly at the transition points.

Other analytical models to represent M–θ relationships of connections are the polynomial model (Frye and Morris 1975), the exponential model (Chen and Lui 1987; Chen and Kishi 1989), the cubic B-spline model (Jones et al. 1980), and the Richard model (Elsati and Richard 1996). In general, the polynomial model gives a better approximation in comparison with linear, bilinear, and piecewise linear models. However, it represents an inherent oscillatory approximation. In some cases, the first derivative of the connection stiffness may be discontinuous or possibly negative, which is physically impossible. Hence, it may provide erroneous estimates of the connection stiffness at several transition points. The exponential model was proposed using laboratory data for monotonic loading. However, since at least six parameters must be evaluated to use this model, it can be very inefficient and very time-consuming. The cubic B-spline model uses a cubic polynomial to fit segments of the M–θ curve. The first and second derivatives are maintained between adjacent segments, using a cubic function. Using this model, good representations of M–θ curves are expected. However, several parameters also need to be evaluated, making the cubic-B-spline model inefficient.

The Richard four-Parameter Model representing the M–θ curves of beam-to-column connections is relatively straightforward and all the parameters have physical meaning indicating that they can be easily interpreted and verified. The four parameters of the model are initial stiffness (k), plastic stiffness (k_P), reference moment (M_0), and curve shape parameter (N). They are shown in Figure 8.1. The model had many advantages including that it is relatively easy to differentiate for the calculation of stiffness and other parameters in the formulation. A computer program was developed using available experimental results with panel zone deformation (Elsati and Richard 1996). The required four parameters (k, k_P, M_0, and N) of the Richard model are estimated using available experimental results. If a connection is defined properly in terms of the sizes of the members and details of the connections (number, types, bolt diameters, etc.), the program will give the most appropriate values of the four parameters fitting the available experimental data. By generating the most appropriate M–θ curves for all the connections, this FEM algorithm has been extensively verified and applied for the deterministic response of steel structures excited by seismic loading

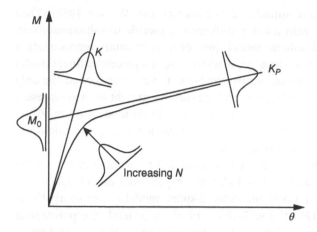

Figure 8.1 $M-\theta$ curve using the Richard four-parameter model.

applied in the time domain considering FR and PR connections (Reyes-Salazar and Haldar 2000, 2001; Mehrabian et al. 2005, 2009). The Richard four-parameter model appears to be the best option to incorporate rigidities of beam-to-column connections. It is practical, beneficial, and suitable to include this model in the computer algorithm presented in Section 8.2.3. The incorporation of this model in the FE algorithm and REDSET is presented next.

8.2.5 Incorporation of Connection Rigidities in the FE Formulation Using Richard Four-Parameter Model

A typical $M-\theta$ curve using the Richard model can be expressed as:

$$M(\theta) = \frac{(k - k_P)\theta}{\left(1 + \left|\frac{(k - k_P)\theta}{M_0}\right|^N\right)^{\frac{1}{N}}} + k_P\theta \tag{8.22}$$

where M is the connection moment and other parameters were defined earlier.

An ordinary beam-column element is used to represent a PR connection in the FE algorithm. However, since the stiffness representing the partial rigidity depends on θ, the stiffness of connection elements needs to be updated at each iteration. It is accomplished by updating Young's modulus (E_c) as:

$$E_c(\theta) = \frac{l_c}{I_c} K_c(\theta) = \frac{l_c}{I_c} \frac{\partial M(\theta)}{\partial \theta} \tag{8.23}$$

where l_c, I_c, and $K_c(\theta)$ are the length, the moment of inertia, and the tangent stiffness of the connection element, respectively.

The term $K_c(\theta)$ in Eq. 8.23 is calculated by taking the first derivative of Eq. 8.22 with respect to θ. It can be represented as:

$$K_c(\theta) = \frac{\partial M(\theta)}{\partial \theta} = \frac{(k - k_P)}{\left(1 + \left|\frac{(k - k_P)\theta}{M_0}\right|^N\right)^{\frac{N+1}{N}}} + k_P \tag{8.24}$$

The Richard four-parameter model represents only the monotonically increasing loading portion of the $M-\theta$ curve. However, the unloading and reloading behavior of PR connections need to be considered for the nonlinear dynamic (seismic, wave, etc.) analyses. The behavior is developed using the Masing rule as suggested by Colson (1991). A general type of Masing model can be defined with a virgin loading curve as:

$$f(M, \theta) = 0 \tag{8.25}$$

and its unloading and reloading curve can be described as:

$$f\left(\frac{M - M_a}{2}, \frac{\theta - \theta_a}{2}\right) = 0 \tag{8.26}$$

where (M_a, θ_a) is the load reversal point as shown in Figure 8.2.

Using the Masing rule and the Richard four-parameter model represented by Eqs. 8.22 and 8.24, the unloading and reloading behavior of a PR connection can be generated as:

$$M(\theta) = M_a - \frac{(k - k_P)(\theta_a - \theta)}{\left(1 + \left|\frac{(k - k_P)(\theta_a - \theta)}{2M_0}\right|^N\right)^{\frac{1}{N}}} - k_P(\theta_a - \theta) \tag{8.27}$$

Figure 8.2 Loading, unloading, and reverse loading model for PR connections.

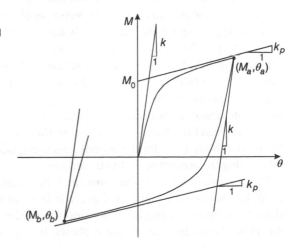

and

$$K_c(\theta) = \frac{\partial M(\theta)}{\partial \theta} = \frac{(k - k_P)}{\left(1 + \left|\frac{(k - k_P)(\theta_a - \theta)}{2M_0}\right|^N\right)^{\frac{N+1}{N}}} + k_P \tag{8.28}$$

If (M_b, θ_b) is the next load reversal point, as shown in Figure 8.2, the reloading relationship between M and θ can be obtained by simply replacing (M_a, θ_a) with (M_b, θ_b) in Eqs. 8.27 and 8.28. Hence, Eqs. 8.22 and 8.24 are used when the connection is loading and Eqs. 8.27 and 8.28 are used when the connection is unloading and reloading. It represents the hysteretic behavior of a PR connection.

In summary, major sources of nonlinearity considered in the formulation are large deformations, material properties, and connection conditions. Both $P-\delta$ and $P-\Delta$ effects are considered for large deformation. Material nonlinearity arises as a result of the constitutive relationship of the material. Considering the usual practice in the profession of steel structures, the material nonlinearity is considered to be elastic-perfectly plastic (Hinton and Owen 1986). Nonlinearity caused by the PR connection condition is also incorporated in the formulation.

8.3 Pre- and Post-Northridge Steel Connections

During the Northridge earthquake of 1994, welds in beam-column connections in several steel frame structures failed in a brittle manner. It was reported that several connections in steel frames also developed similar cracks in the welds during the Loma Prieta earthquake of 1989. Beam-column connections in steel frames are generally represented as FR type. However, as discussed in the previous section, from the experimental investigations, it was reported that these connections are rarely FR type (Mehrabian et al. 2005; Reyes-Salazar and Haldar 2001). They are PR with different rigidities. Past studies indicated that representing beam-column connections as PR type changes the structural dynamic properties (stiffness, damping, frequency, mode shape, etc.) (Reyes-Salazar and Haldar 2001). Structural responses are expected to be very different if appropriate connection conditions are incorporated in the formulation significantly altering the serviceability LSFs for the overall lateral displacement and inter-story drifts influencing the estimation of the underlying risk.

The profession expects that structural behavior, including the connection condition, needs to be considered as realistically as practicable when estimating risk or reliability. To amplify this point, a typical connection that failed during the 1994 Northridge earthquake is shown in Figure 8.3. It will be referred to

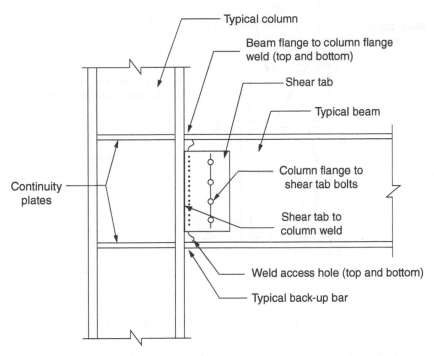

Typical column

Beam flange to column flange weld (top and bottom)

Shear tab

Typical beam

Column flange to shear tab bolts

Continuity plates

Shear tab to column weld

Weld access hole (top and bottom)

Typical back-up bar

Figure 8.3 Typical pre-Northridge connection. *Source:* Mehrabian et al. (2005).

hereafter as the pre-Northridge connection. After an extensive study, the Structural Engineering Association of California (SEAOC) recommended against using these connections. Subsequently, several improved post-Northridge connections were proposed to provide more ductility and increase the energy absorption capacity during seismic excitations.

Some of the post-Northridge connections reported in the literature are cover-plated (Engelhardt and Sabol 1998), spliced beam (Popov et al. 1998), sided-plated, bottom haunch, connections with vertical ribs, connections with reduced beam sections (RBS) or dog-boned, and slotted-web beam-column connections (Richard et al. 1997). Seismic Structural Design Associates (SSDA) proposed a unique proprietary slotted-web (Slotted-Web) moment connection by cutting slots in the web of the beam as shown in Figure 8.4. They tested several full-scale beam-column connection models using the ATC-24 test protocol (Richard et al. 1997) and shared the test data with the authors. The test results clearly indicated that the slots in the web of the beam introduced the desirable behavior without reducing the initial stiffness of the connection. However, the presence of two slots on the web raised concern from some scholars and practitioners.

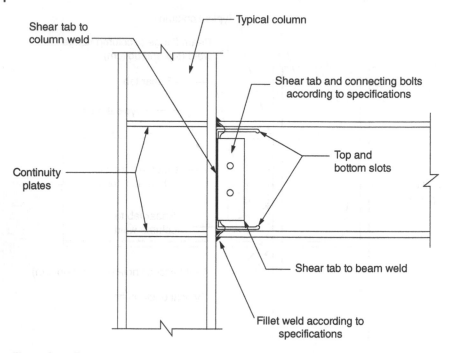

Figure 8.4 Typical post-Northridge connection. *Source:* Mehrabian et al. (2005).

The authors' team analytically confirmed the significantly improved behavior observed during testing (Mehrabian and Haldar 2005, 2007; Mehrabian et al. 2005) without producing any adverse effect predicted by the critics. They extended the test results to consider test configurations not considered by SSDA. The improved behavior of the structure in the presence of these connections indicates that the underlying risk was reduced. An acceptable reliability estimation method should be able to document the reduction in the estimated risk with the help of appropriate physics-based modeling of all connections. This observation clearly indicates that a physics-based formulation is a necessity; it should not be an optional or desirable item. It will be discussed in more detail with examples later.

By using actual test data obtained by SSDA, Mehrabian et al. (2005) estimated the four parameters of the Richard curve. For the same connection without the slots (pre-Northridge connection), four Richard parameters were also estimated. Obviously, the parameters for the pre-Northridge and post-Northridge connections were very different. They also need to be considered as PR connections with different rigidities.

8.4 Performance-Based Seismic Design

The PBSD concept is relatively new but expected to be incorporated into future design codes and guidelines, at least in the United States. Because of its newness, additional discussions are necessary to introduce the concept.

8.4.1 Background Information and Motivation

The enormous amount of property damages caused by the 1994 Northridge Earthquake prompted the profession to find an alternative to the currently accepted design criterion of LS. Although LS was not compromised during this earthquake, the structural damage caused by it was enormous. It indicated a major deficiency that existed at that time in the prescriptive design guidelines to protect life. To address the related issues, under the financial sponsorship of the Federal Emergency Management Agency (FEMA), a major study was initiated by SAC [a joint venture of the SEAOC, Applied Technology Council (ATC), and California Universities for Research in Earthquake Engineering (CUREE)]. The main objective of the joint venture was to develop recommendations for a more robust design and construction of steel structures and to propose alternative design criteria to avoid adverse economic consequences.

In 1995, SEAOC published a report titled Performance-Based Seismic Engineering of Buildings (SEAOC Vision 1995). They published another report in 2000. These publications outlined a set of recommendations for designing buildings with predictable and predefined seismic performance levels. FEMA published another document in 1997 suggesting guidelines for the seismic rehabilitation of buildings (FEMA-273 1997). In 2000, SAC published major findings in a series of reports (FEMA-350 2000; FEMA-351 2000; FEMA-352 2000; FEMA-353 2000; FEMA-354 2000; FEMA-355C 2000; FEMA-355F 2000). These reports and guidelines provide the foundation for the PBSD concept for structures for implementation in building design codes.

The seismic loading is very unpredictable at a particular site and the estimation of dynamic responses of a structure in the presence of many other sources of uncertainty can be very challenging. Since the sources of uncertainty cannot be completely eliminated in the seismic design, the safety of structures cannot be assured. SAC advocated for the managing of the underlying risk. To achieve this objective, SAC proposed to correlate underlying risk with different performance levels and suggested that the designers and/or owners would decide if the underlying risk of a structure already designed was acceptable to them or not (FEMA-350 2000). The report defined the performance level as "the intended post-earthquake condition of a building; a well-defined point on a scale measuring how much loss is caused by earthquake damage." Performance levels are defined

| Immediate occupancy | Life safety | Collapse prevention |

Figure 8.5 Different performance levels. *Source:* Adapted from FEMA-350 (2000).

as immediate occupancy (IO) (a performance level of very low structural damage), LS, and collapse prevention (CP) (a state of extreme structural damage). They are conceptually shown in Figure 8.5. Unfortunately, SAC did not recommend any specific procedure for risk estimation.

The available literature on PBSD is limited. FEMA-355F (2000) identified the following six important objectives that need to be incorporated into the formulation. They are: (i) account for uncertainty in the performance associated with unanticipated events, (ii) set realistic expectations for performance, (iii) assess performance variables in similar buildings located nearby, (iv) develop a reliability framework, (v) set representative performance levels for various seismic hazards, and (vi) quantify local and global structural behaviors leading to collapse.

To satisfy these objectives, PBSD proposed a set of procedures by which a structural system should be designed in a controlled manner. The concept can be implemented by following five sequential steps: (i) select performance objectives, (ii) develop a preliminary design, (iii) assess performance capability, (iv) check performance capability with allowable values, in terms of the associated risks, and (v) if risks are not acceptable, revise the initial design. These steps are novel and futuristic in nature and can be used as an alternative to the current practice of protecting life.

8.4.2 Professional Perception of PBSD

PBSD is an evolving methodology that has become increasingly popular among professional engineers, particularly in the seismically active regions of California. However, there remain several hurdles that need to be addressed. Some of the major challenges are related to the development of precise definitions of performance objectives that will be accepted universally, quantify performance levels, and estimate the underlying risk of applying seismic loading in the time domain satisfying the specific performance level. The recent drive to use PBSD for the design of new as well as existing building structures has led to rapid developments in analytical approaches (Hamburger and Hooper 2011). The basic idea of PBSD is to design a structure so that it will satisfy various performance levels or criteria when subjected to the corresponding different earthquake ground motions. From the perspective of professionals, PBSD has two principal objectives. The first one is to couple the structural requirements with performance expectations to guarantee that hazards are treated consistently. The other is to assure financial losses associated with damages correlate well with the expectations. However, one of the major challenges in implementing PBSD is the unavailability of the computational tools required for the estimation of the seismic risks corresponding to the different performance levels (Ghobarah 2001).

There are several important issues in implementing PBSD. The selection of performance levels and the appropriate mathematical models to represent structural behavior could be very complicated. When seismic loading is applied in the time domain, the structure is expected to develop various sources of nonlinearities and the mathematical model to capture such behavior can be very demanding. To study such nonlinear behavior, the structure is generally represented by finite elements (FEs). Considering accuracy and efficiency, there may not be any uniformity in representing real, large, and complicated structural systems by FEs and, in fact, it can be very difficult. To capture the dynamic application of the seismic loading, several methods with various degrees of sophistication are suggested in the current design guidelines, including pseudo-static to time domain application of the excitation (ASCE/SEI 7-10 2010). The most sophisticated analysis will require a structure to be represented by nonlinear FEs and the dynamic seismic loading must be applied in the time domain. The proper application of seismic loading in the time domain depends on the selection of ground motions. Commonly used design guidelines recommend the use of seven or more ground motions applied in the time domain for the structural response analyses (ASCE/SEI 7-10 2010). Ground motions can be recorded (real) or simulated (artificial), depending on the availability of the records. The main challenge in the profession remains is the explicit quantification of the vulnerability (reliability) of structures to future design seismic loading. In addition, commonly used analytical models of nonlinear steel structures are too idealized and may not satisfy the underlying physics. As discussed earlier, beam-to-column connections introduce nonlinearity even when the seismic loading is very small. The structural dynamic properties (stiffness, damping, frequency, mode shape, etc.) are expected to be quite different if the connection conditions are modeled realistically. The major point of this discussion is that all desirable performance-enhancing features introduced in the design of a structure must be appropriately incorporated in developing PBSD.

The most important knowledge gap in implementing PBSD is the risk evaluation, considering all major sources of nonlinearity and uncertainty, and applying the seismic loading in the time domain. In the context of PBSD, the performance requirements are suggested by the owner and/or engineers according to the risk they are willing to take and accept the corresponding economic consequences. PBSD is essentially a very sophisticated risk-based design concept. However, there is no reliability analysis procedure currently available that will be acceptable to all concerned parties. It is important at this stage to explore if REDSET can fill this knowledge gap and help to design steel structures using PBSD in seismically active regions of the world. PBSD reflects a major paradigm shift in seismic design moving away from the prescriptive codified approach of LS practiced in the past (Hamburger et al. 2009).

8.4.3 Building Codes, Recommendations, and Guidelines

Currently, the two most important codes for the structural design used in the United States are the *Minimum Design Loads for Buildings and Other Structures* (ASCE/SEI 7-10 2010), and the *International Building Code* (IBC 2000, 2012, 2018). These building codes are prescriptive in nature and fail to guarantee the real seismic performance of structures satisfying the intent of PBSD (Hamburger and Hooper 2011). As new information is being generated on PBSD, several guidelines and recommendations were reported in the literature incorporating the new information. Some of them are: *An Alternative Procedure for Seismic Analysis and Design of Tall Buildings Located in the Los Angeles Region* (LATBSDC 2011), *Guidelines for Performance-Based Seismic Design of Tall Buildings* (TBI 2010), *Seismic Provisions for Structural Steel Buildings* (AISC 341-10 2010), *Seismic Performance Assessment of Buildings* (FEMA P-58 2012), *NEHRP Recommended Seismic Provisions: Design Examples* (FEMA P-751 2012), and *Seismic Evaluation and Retrofit of Existing Buildings* (ASCE/SEI 41-17 2017). As mentioned earlier, FEMA with the help of the ATC, the Building Seismic Safety Council (BSSC) and the American Society of Civil Engineers (ASCE) provided leadership in developing necessary guidelines for PBSD specifically for steel structures in the United States. The codes, design guidelines, and recommendations reflect a considerable advancement in PBSD. However, they did not specify any acceptable method to estimate the underlying seismic risk although PBSD is a very sophisticated risk-based design concept. The role of REDSET in implementing PBSD is explored and documented in the following sections.

8.4.4 Performance Levels

FEMA-350 (2000) recommends three performance levels for PBSD for steel structures: IO, LS, and CP, as conceptually shown in Figure 8.5. To avoid any ambiguity in describing different performance levels pictorially, they are presented in a described way in Table 8.1.

Table 8.1 Performance levels based on PBSD.

Performance level	Description
IO	The structure has sustained minimal or no damage to its structural elements and only minor damage to its nonstructural components.
LS	Represents a state of extensive damage to structural and nonstructural components. While the risk to life is low, repairs may be required before re-occupancy can occur.
CP	The building has reached a state of impeding partial or total collapse.

8.4.5 Target Reliability Requirements to Satisfy Different Performance Levels

One critical issue to implement PBSD is the target or acceptable reliability requirements corresponding to specific performance levels. However, they still did not suggest acceptable underlying reliability levels. FEMA-350 (2000) proposed performance requirements for IO, LS, and CP in terms of probability of exceedance (PE), earthquake return period, and allowable overall lateral and/or inter-story drifts. They are summarized in Table 8.2. The term H in Table 8.2 represents the total and inter-story height of the building to consider the overall and inter-story drifts, respectively, to develop the corresponding LSFs of interest.

These requirements relate to the serviceability LSFs. Since in most seismic designs of steel structures, the overall and inter-story drift requirements are more critical than the performance in the strength of the properly designed structural elements using recommended procedures, these recommendations for PBSD are acceptable. The intent is to relate the performance levels with the corresponding risk so that the owner and/or engineer are aware of the underlying risk. If they are unwilling to accept the risk, the situation can be mitigated in several ways including buying insurance to cover the additional risk.

Table 8.2 Structural correlation of performance levels.

Performance level	Probability of exceedance (PE)	Earthquake return period	Allowable drift
IO	50% in 50 years	72-year	0.007^*H
LS	10% in 50 years	475-year	0.025^*H
CP	2% in 50 years	2475-year	0.050^*H

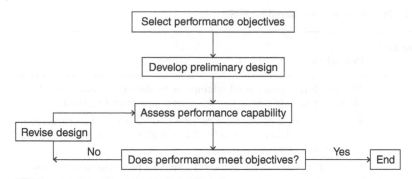

Figure 8.6 Elements of PBSD.

8.4.6 Elements of PBSD and Their Sequences

In summary, PBSD is a sophisticated risk-based design concept and needs to be implemented sequentially in a systematic way. All the five basic steps identified in (ASCE/SEI 41-17 2017) are shown in Figure 8.6. They are: (i) select performance objectives, (ii) develop a preliminary design, (iii) assess performance capability, (iv) check if performance objectives are met, and (v) revise the design if performance objectives are not met. They need to be carried out sequentially as shown in Figure 8.6.

8.4.7 Explore Suitability of REDSET in Implementing PBSD

At the risk of repetition, it is important to state that REDSET is primarily developed for estimating the underlying risk of CDNES excited in the time domain. To consider uncertainty in the dynamic excitation applied in the time domain, a suite of earthquake time histories needs to be used as suggested in the design code. At present, PBSD is essentially developed for steel structures to satisfy the serviceability-related LSFs of overall lateral and inter-story drifts. For realistic structural systems, these LSFs are implicit in nature. In developing REDSET in Chapter 5, it would approximately generate explicit expressions for the implicit LSFs using a second-order polynomial representing an IRS in the failure region considering all major sources of uncertainty in the resistance- and gravity load-related RVs. At least nine different advanced experimental sampling design schemes were proposed to generate an IRS. Considering efficiency, accuracy, and robustness, Sampling Scheme MS3-KM appears to be the most attractive and will be used to implement PBSD. For developing the required IRS, the structural response data will be generated by conducting a few deterministic nonlinear FE analyses of a structure excited by the multiple earthquake acceleration time histories. The acronym TNDFEA was used to describe *total number of dynamic finite element analyses* of the structure. To develop the required LSFs to implement PBSD, the information on the permissible or allowable lateral defection δ_{allow} will be used as summarized in Table 8.2 corresponding to the required performance level. Once the required LSFs are explicitly available, the information on the reliability index β and the corresponding probability of failure p_f will be extracted using FORM. Obviously, the procedure needs to be repeated for all the selected earthquake time histories one by one. It is now necessary to showcase how REDSET can be used to implement PBSD.

8.5 Showcasing the Implementation of PBSD

As mentioned in the previous chapters, before initiating any risk evaluation study, it is essential to identify all the RVs in the formulation and then quantify the uncertainty in them. And when necessary, verify all the necessary steps by applying the algorithm to a relatively small problem excited by a known earthquake acceleration time history. After satisfactory verification, CNDES can be considered. They need to be excited by multiple design earthquake time histories applied in the time domain to generate the response data following a sampling design scheme to obtain an explicit expression for an IRS and the corresponding LSF. The same steps are taken in this chapter also.

To meet these objectives, four numerical examples are considered. In the first example, a 2-story steel frame excited by the three acceleration time histories recorded during the 1994 Northridge earthquake is considered. The beam-column connections are considered to be FR (pre-Northridge type) and PR (post-Northridge type) in this chapter to study the implications of flexibility in the connection conditions. The structure and the connections were represented by the assumed stress-based FEM. The information on risk is extracted using REDSET with sampling Scheme MS3-KM and then verified using the basic MCS technique. After the successful verification, three additional steel structures of 3-, 9-, and 20-story buildings designed by experts satisfying post-Northridge design guidelines, reported by FEMA are considered. These buildings were also considered in Chapter 6 to verify REDSET but they were represented by the displacement-based FEM. The major difference between Chapter 6 and this chapter is instead of displacement-based FEM, the stress-based FEM is used here. In addition, the beam-column connections are considered to be FR (pre-Northridge) and PR (post-Northridge) types as in the previous example. They are excited by the earthquake time histories required for the IO, LS, and CP performance levels and their risks are estimated using the corresponding allowable drift values suggested for the PBSD guidelines.

8.5.1 Verification of REDSET – Reliability Estimation of a 2-Story Steel Frame

To verify the implementation potential of REDSET with Sampling Scheme MS3-KM for PBSD using the assumed stress-based FEM, a 2-story steel frame shown in Figure 8.7 is considered. This example was also considered by Huh (1999) to verify the SFEM discussed earlier. It is a relatively small structure and is considered the first to check the capability of REDSET in implementing PBSD by comparing the results with the basic MCS. The frame is excited by the three time histories recorded during the 1994 Northridge earthquake, as shown in Figure 8.8.

For ease of discussion, initially, a typical connection is represented by the pre-Northridge connection as shown in Figure 8.3, although SEAOC recommended not to use it. As an alternative, all the connections are also considered to be post-Northridge-type PR connections with two slots in the web of the beam, as shown in Figure 8.4. The four parameters (k, k_P, M_0, and N) of the Richard model to define both types of PR connection are estimated by the available computer program. The information is summarized in Table 8.3.

The uncertainties associated with resistance-related parameters are widely reported in the literature (Ellingwood et al. 1980; Haldar and Mahadevan 2000a, 2000b; Nowak and Collins 2012) and used in the study for steel structures to study the implementation of PBSD. All structural elements are made with W-sections as reported in AISC (2011) design manual. Statistical characteristics of RVs are expressed in terms of distribution, mean to nominal ratio ($\overline{X}/X_N$),

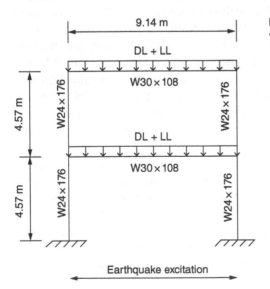

Figure 8.7 A 2-story frame used for validations.

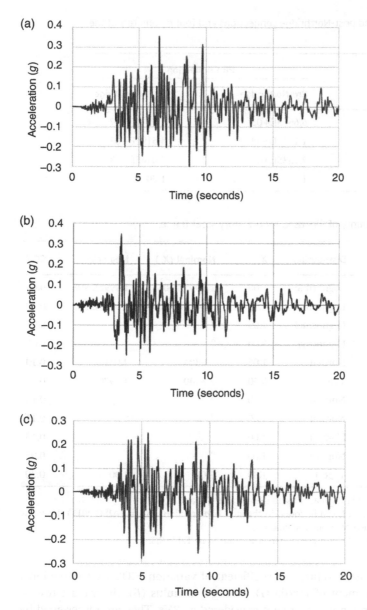

Figure 8.8 Acceleration time histories recorded at (a) Canoga park station time history; (b) Nordhoff fire station; and (c) Roscoe Blvd. station.

Table 8.3 Pre- and post-Northridge connections and four parameters of the Richard Model.

	PR Connection	
Parameter	Pre-Northridge	Post-Northridge
k (kN-m/rad)	6.6663E+05	1.9546E+07
k_P (kN-m/rad)	1.1113E+04	4.5194E+03
M_0 (kN-m)	7.7835E+02	2.0145E+03
N	1.10	1.00

Table 8.4 Uncertainty of RVs used for a 2-story steel frame.

Random variable	Distribution	$\overline{X}/X_N$	Nominal (X_N)	Mean ($\overline{X}$)	COV
E (kN/m^2)	Lognormal	1.00	1.9995E+08	1.9995E+08	0.06
Fy (kN/m^2)	Lognormal	1.05	3.4474E+05	3.6197E+05	0.10
A (m^2)	Lognormal	1.00	a	b	0.05
I_x (m^2)	Lognormal	1.00	a	b	0.05
DL (kN/m^2)	Normal	1.05	3.8304	4.0219	0.10
LL (kN/m^2)	Type 1	0.50	2.3940	1.1970	0.25
k (kN-m/rad)	Normal	1.00	a	c	0.15
kp (kN-m/rad)	Normal	1.00	a	c	0.15
M_0 (kN-m)	Normal	1.00	a	c	0.15
N	Normal	1.00	a	c	0.05
g_e	Type 1	1.00	1.00	1.00	0.20

[a] Nominal value (X_N) is calculated using mean value ($\overline{X}$) and $\overline{X}/X_N$.
[b] Mean values of A and I_x can be found in the steel construction manual (AISC 2011). They are considered RVs for every girder and column.
[c] Mean values of four Richard parameters are reported in Table 8.3.

nominal value, mean value, and coefficient of variation (COV). Cross-sectional area (A), the moment of inertia (I), Young's modulus (E), the yield stress of columns (Fy_c) and girders (Fy_g) are considered as RVs. They are represented by lognormal distribution with mean values as given in the AISC design manual with the COVs of 0.05, 0.05, 0.06, 0.10, and 0.10, respectively. The information is summarized in Table 8.4. The four parameters of the Richard models to represent PR connections are considered to be normally distributed with the mean values given in Table 8.3 with COVs given in Table 8.4.

As mentioned earlier, to showcase PBSD guidelines, only the serviceability LSFs of the overall and the inter-story drifts for steel structures need to be considered. They can be represented as:

$$g(\mathbf{X}) = \delta_{\text{allow}} - y_{\max}(\mathbf{X}) = \delta_{\text{allow}} - \hat{g}(\mathbf{X}) \tag{8.29}$$

where δ_{allow} is the permissible or allowable value to satisfy a specific performance level, $\hat{g}(\mathbf{X})$ is the predicted second-order polynomial representing the IRS of interest, and $\mathbf{X}$ is a vector representing all the design RVs. As mentioned earlier, the IRSs are generated by fitting the response data obtained by the stress-based FEM and REDSET Sampling Scheme MS3-KM. Their allowable or recommended δ_{allow} values for the two LSFs are suggested in FEMA-350 (2000) as summarized in Table 8.2. The allowable overall lateral and inter-story drifts are considered to be 2.86 cm and 1.43 cm, respectively for the IO performance level. Since the maximum displacement is estimated by the nonlinear time history analysis, the permissible values are increased by 125% according to ASCE/SEI 7–10 (ASCE 2010).

The total and reduced number of significant RVs for the problem are found to be $k = 14$ and $k_r = 5$, respectively. The joints are considered to be FR- and PR-type with pre-Northridge connections, and PR-type with post-Northridge connections with slotted holes. The reliability index β and the corresponding probability of failure p_f are estimated using REDSET Sampling Scheme MS3-KM. The results are summarized in Table 8.5. The results are verified using 50 000 cycles of MCS. For this example, TNDFEA required to implement REDSET is 94.

Several important observations can be made from the results given in the table. The β and the corresponding p_f values are essentially the same for REDSET using TNDFEA of 94 and 50 000 MCS. The results confirm that the REDSET algorithm can be used to implement PBSD. It is very interesting to observe that the behavior of the frame in the presence of FR and post-Northridge PR connections are also essentially the same for all three ground motions confirming the observations made during the experiments that the slots in the web do not alter the expected behavior even when connections are considered to be PR type. The p_f values for the frame in the presence of pre-Northridge connections are at least one order of magnitude higher than that of post-Northridge connections in some cases indicating the frame is much weaker and SEAOC was correct in recommending not to use them. Furthermore, both p_f and β are different for the three ground motions, indicating the difference in the frequency contents of them. The use of multiple time histories is essential to address the issue as indicated in the recent design guidelines. The results indicated that the recommendations suggested in post-Northridge design guidelines are steps in the right direction and REDSET is capable of conclusively confirming these observations.

Table 8.5 Reliability results for a 2-story steel frame.

Ground motion	LSF	Method	FR		PR (pre-Northridge)		PR (post-Northridge)	
			β (TNDFEA)	p_f	β (TNDFEA)	p_f	β (TNDFEA)	p_f
Canoga park station	Overall drift	REDSET	3.5232 (94)	0.000213	3.4063 (94)	0.000329	3.6208 (94)	0.000147
		MCS	3.5149 (50 000)	0.000220	3.4141 (50 000)	0.000320	3.6331 (50 000)	0.000140
	Inter-story drift	REDSET	3.2429 (94)	0.000592	2.9825 (94)	0.001400	3.3357 (94)	0.000425
		MCS	3.2389 (50 000)	0.000600	2.9845 (50 000)	0.001420	3.3139 (50 000)	0.000460
Nordhoff fire station	Overall drift	REDSET	3.6949 (94)	0.000110	3.1690 (94)	0.000765	3.8853 (94)	0.000051
		MCS	3.7190 (50 000)	0.000100	3.1708 (50 000)	0.000760	3.8461 (50 000)	0.000060
	Inter-story drift	REDSET	3.2954 (94)	0.000491	2.8544 (94)	0.002200	3.5170 (94)	0.000218
		MCS	3.3139 (50 000)	0.000460	2.8627 (50 000)	0.002100	3.5149 (50 000)	0.000220
Roscoe Blvd. station	Overall drift	REDSET	3.6508 (94)	0.000131	3.0009 (94)	0.001300	3.9039 (94)	0.000040
		MCS	3.6331 (50 000)	0.000140	3.0068 (50 000)	0.001320	4.1075 (50 000)	0.000020
	Inter-story drift	REDSET	3.2528 (94)	0.000571	2.8192 (94)	0.002400	3.5969 (94)	0.000161
		MCS	3.2585 (50 000)	0.000560	2.8149 (50 000)	0.002440	3.5985 (50 000)	0.000160

8.6 Implementation Potential of PBSD – 3-, 9-, and 20-Story Steel Buildings

After successfully verifying REDSET using a relatively small problem, it is now necessary to document or showcase how to implement the PBSD procedure for the design of realistic CNDES.

8.6.1 Description of the Three Buildings

To meet the objective of implementing PBSD using REDSET, 3-, 9-, and 20-story steel buildings designed by experts in the Los Angeles area are considered, as reported in FEMA-355C (2000). Plans and elevations of these structures are shown in Figures 8.9–8.11. As discussed in Section 6.4, the experts suggested several design options. One option is considered here. They were designed for sites with a stiff soil condition, Soil Class D. The sizes of members are given in the figures. These frames are supposed to be considered benchmark designs. They are expected to be used in any research on PBSD.

8.6.2 Post-Northridge PR Connections

Since SEAOC recommended not to use pre-Northridge PR connections anymore, they are not considered in the following discussions. The statistical characteristics of the four parameters (k, k_P, M_0, and N) of the Richard model for the post-Northridge connections are summarized in Table 8.6. Their nominal values are estimated using the software available to the authors.

8.6.3 Quantification of Uncertainties in Resistance-Related Variables

As discussed in the previous example, uncertainties associated with all resistance-related RVs are collected from the available literature and summarized in Table 8.7.

8.6.4 Uncertainties in Gravity Loads

In most design guidelines (ASCE/SEI 7-10 2010; IBC 2012), gravity loads are classified as dead load (DL) and live load (LL). The uncertainties associated with them are available in the literature (Ellingwood et al. 1980; Haldar and Mahadevan 2000a, 2000b; Nowak and Collins 2012) and used in this study also. DL and LL are represented by the normal and Type 1 distributions with COV of 0.10 and 0.25, respectively. The information is also summarized in Table 8.7.

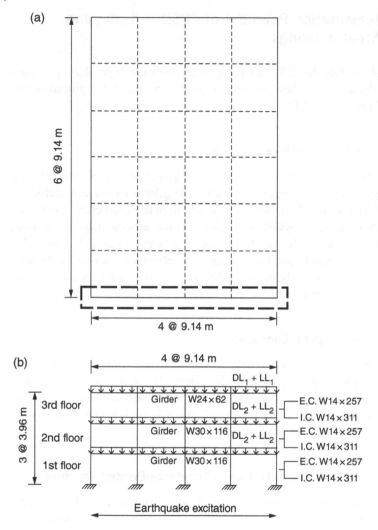

Figure 8.9 A 3-story building: (a) Plan view; (b) Elevation (EC = Exterior columns; IC = Interior columns).

8.6.5 Uncertainties in PR Beam-to-Column Connections

The four parameters k, k_P, M_0, and N are considered normal variables. Their mean values are estimated using the information on the size of the members and details of the connection with the help of a computer program discussed earlier. The values generated by the available computer program are considered mean values.

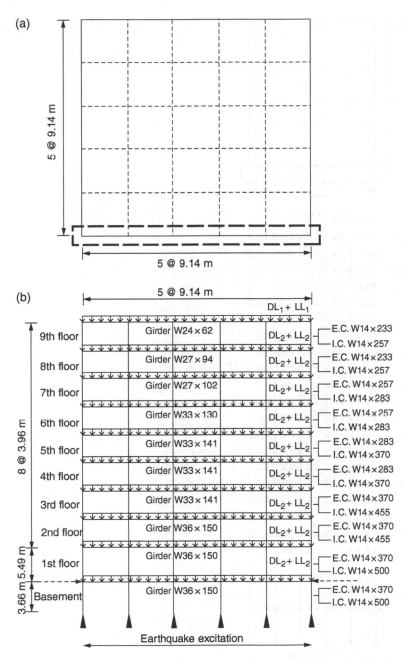

Figure 8.10 A 9-story building: (a) Plan view; (b) Elevation (EC = Exterior columns; IC = Interior columns).

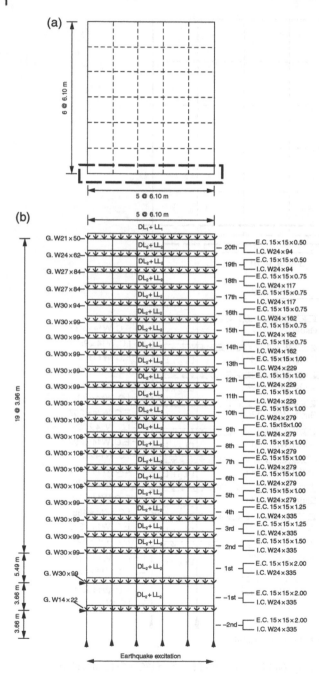

Figure 8.11 A 20-story building: (a) Plan view; (b) Elevation (EC = Exterior columns; IC = Interior columns; G = Girder).

Table 8.6 Four parameters of Richard model for the 3-, 9-, and 20-story buildings considering post-Northridge PR beam-to-column connections.

Model	Level[a]	Girder	k (kN-m/rad)	kp (kN-m/rad)	M_0 (kN-m)	N
				Four Richard parameters		
3-story	Roof	W24×62	8.6433E+06	4.5194E+03	8.1519E+02	1.00
	3	W30×116	2.1354E+07	4.5194E+03	2.2134E+03	1.00
	2	W30×116	2.1354E+07	4.5194E+03	2.2134E+03	1.00
9-story	Roof	W24×62	8.6433E+06	4.5194E+03	8.1519E+02	1.00
	9	W27×94	1.5705E+07	4.5194E+03	1.5920E+03	1.00
	8	W27×102	1.7230E+07	4.5194E+03	1.7597E+03	1.00
	7	W33×130	2.6382E+07	4.5194E+03	2.7664E+03	1.00
	6	W33×141	2.9037E+07	4.5194E+03	3.0585E+03	1.00
	5	W33×141	2.9037E+07	4.5194E+03	3.0585E+03	1.00
	4	W33×141	2.9037E+07	4.5194E+03	3.0585E+03	1.00
	3	W36×150	3.2822E+07	4.5194E+03	3.4748E+03	1.00
	2	W36×150	3.2822E+07	4.5194E+03	3.4748E+03	1.00
	1	W36×150	3.2822E+07	4.5194E+03	3.4748E+03	1.00
20-story	Roof	W21×50	6.2142E+06	4.5194E+03	5.4798E+02	1.00
	20	W24×62	8.6433E+06	4.5194E+03	8.1519E+02	1.00
	19	W27×84	1.3784E+07	4.5194E+03	1.3807E+03	1.00
	18	W27×84	1.3784E+07	4.5194E+03	1.3807E+03	1.00
	17	W30×99	1.7626E+07	4.5194E+03	1.8032E+03	1.00
	16	W30×99	1.7626E+07	4.5194E+03	1.8032E+03	1.00
	15	W30×99	1.7626E+07	4.5194E+03	1.8032E+03	1.00
	14	W30×99	1.7626E+07	4.5194E+03	1.8032E+03	1.00
	13	W30×99	1.7626E+07	4.5194E+03	1.8032E+03	1.00
	12	W30×99	1.7626E+07	4.5194E+03	1.8032E+03	1.00
	11	W30×108	1.9546E+07	4.5194E+03	2.0145E+03	1.00
	10	W30×108	1.9546E+07	4.5194E+03	2.0145E+03	1.00
	9	W30×108	1.9546E+07	4.5194E+03	2.0145E+03	1.00
	8	W30×108	1.9546E+07	4.5194E+03	2.0145E+03	1.00
	7	W30×108	1.9546E+07	4.5194E+03	2.0145E+03	1.00
	6	W30×108	1.9546E+07	4.5194E+03	2.0145E+03	1.00
	5	W30×99	1.7626E+07	4.5194E+03	1.8032E+03	1.00

(Continued)

Table 8.6 (Continued)

Model	Level[a]	Girder	k (kN-m/rad)	kp (kN-m/rad)	M₀ (kN-m)	N
				Four Richard parameters		
	4	W30×99	1.7626E+07	4.5194E+03	1.8032E+03	1.00
	3	W30×99	1.7626E+07	4.5194E+03	1.8032E+03	1.00
	2	W30×99	1.7626E+07	4.5194E+03	1.8032E+03	1.00
	1	W30×99	1.7626E+07	4.5194E+03	1.8032E+03	1.00
	−1	W14×22	1.8755E+06	4.5194E+03	7.0729E+01	1.00

[a] See Figures 8.9–8.11 for more details.

Table 8.7 Uncertainty of RVs used for 3-, 9-, and 20-story steel frames.

Random variable	Distribution	$\overline{X}/X_N$	Nominal (X_N)	Mean ($\overline{X}$)	COV
E (kN/m²)	Lognormal	1.00	1.9995E+08	1.9995E+08	0.06
Fy_G (kN/m²)[a]	Lognormal	1.35	2.4821E+05	3.3509E+05	0.10
Fy_C (kN/m²)[a]	Lognormal	1.15	3.4474E+05	3.9645E+05	0.10
A (m²)	Lognormal	1.00	b	c	0.05
I_x (m²)	Lognormal	1.00	b	c	0.05
DL_R (kN/m²)	Normal	1.05	3.9740	4.1727	0.10
DL_F (kN/m²)	Normal	1.05	4.5965	4.8263	0.10
LL_R (kN/m²)	Type 1	0.40	2.3940	0.9576	0.25
LL_F (kN/m²)	Type 1	0.40	2.3940	0.9576	0.25
k (kN-m/rad)	Normal	1.00	b	d	0.15
kp (kN-m/rad)	Normal	1.00	b	d	0.15
M_0 (kN-m)	Normal	1.00	b	d	0.15
N	Normal	1.00	b	d	0.05
g_e	Type 1	1.00	1.00	1.00	0.20

[a] Yield stress of girder or column cross-section reported in FEMA-355C (2000).
[b] Nominal value (X_N) is calculated using mean value ($\overline{X}$) and $\overline{X}/X_N$.
[c] Mean values of A and I_x can be found in the steel construction manual (AISC 2011). They are considered RVs for every girder and column.
[d] Mean values of four Richard parameters are reported in Table 8.6.

The COVs of k, k_P, M_0, and N are considered to be 0.15, 0.15, 0.15, and 0.05, respectively. The information is also summarized in Table 8.7.

8.6.6 Uncertainties in Seismic Loading

As discussed in Chapter 4, multiple earthquake acceleration time histories are considered to address uncertainties associated with seismic loading. In Chapter 4, Alternative 1 – by scaling PEER database, discussed in detail in Section 4.3, was used. For this example, Alternative 2 – using a broadband platform (BBP), discussed in detail in Section 4.5, is used. For PBSD, multiple performance levels must be considered and the earthquake time histories need to be appropriately selected for a specific performance level.

As mentioned in Chapter 4, BPP requires a considerable amount of information on site-specific seismic activities. The information may not be available in regions outside the United States. The use of Alternative 1, i.e. scaling of earthquake time histories available at PEER database will be more appropriate for those cases. In any case, the site-specific target probabilistic ground response spectrum (TPGRS) must be available for both alternatives. Multiple time histories can be generated by fitting the procedures suggested in Section 4.3.

Using the information available for the Los Angeles area where the three buildings are located with Soil Class D (no information on soil is available), Somerville (1997) suggested three sets of earthquake time histories corresponding to return periods of 2475-year (2% PE in 50 years), 475-year (10% PE in 50 years), and 72-year (50% PE in 50 years). They matched the three performance levels suggested in developing the PBSD guidelines. For each performance level, ten ground motions with two horizontal orthogonal components (North-South [N-S] and East-West [E-W]) were suggested, generating twenty ground motions per suite. Somerville (1997) applied scale factors (SFs) to match the TPGRS values, for several periods at 0.3, 1.0, 1.0, and 4.0 seconds, on average, considering Soil Class D. Then, ground motion records were selected based on an approximate deaggregation of the hazard of the zone for each PE level. Preference was given to ground motions recorded in the region. For the ground motions associated with a 2% PE in 50 years, several simulations were conducted using the BBP following rigorous theoretical and computational seismology. Relevant information on each suite of ground motions is summarized in Tables 8.8–8.10 for the performance levels of CP, LS, and IO, respectively. The information on the official name of the earthquakes, SF, peak ground acceleration (PGA), magnitude (M_w), hypocentral distance (R), and duration is listed in tables. Since the information on multiple time histories on a specific site condition for the three performance levels is readily available, it is used here. Implementation of PBSD for different soil conditions will be discussed in Section 8.8.

Table 8.8 Ground motions corresponding to 2% PE in 50 years and CP performance level, Site Class S_D.

EQ	Name	M_w	SF	PGA (g)	Time (sec)
1	1995 Kobe	6.9	1.15	1.282	25.0
2	1995 Kobe	6.9	1.15	0.920	25.0
3	1989 Loma Prieta	7.0	0.82	0.418	20.0
4	1989 Loma Prieta	7.0	0.82	0.473	20.0
5	1994 Northridge	6.7	1.29	0.868	14.0
6	1994 Northridge	6.7	1.29	0.943	14.0
7	1994 Northridge	6.7	1.61	0.926	15.0
8	1994 Northridge	6.7	1.61	1.329	15.0
9	1974 Tabas	7.4	1.08	0.808	25.0
10	1974 Tabas	7.4	1.08	0.991	25.0
11	Elysian Park (simulated)	7.1	1.43	1.295	18.0
12	Elysian Park (simulated)	7.1	1.43	1.186	18.0
13	Elysian Park (simulated)	7.1	0.97	0.782	18.0
14	Elysian Park (simulated)	7.1	0.97	0.680	18.0
15	Elysian Park (simulated)	7.1	1.10	0.991	18.0
16	Elysian Park (simulated)	7.1	1.10	1.100	18.0
17	Palos Verdes (simulated)	7.1	0.90	0.711	25.0
18	Palos Verdes (simulated)	7.1	0.90	0.776	25.0
19	Palos Verdes (simulated)	7.1	0.88	0.500	25.0
20	Palos Verdes (simulated)	7.1	0.88	0.625	25.0

8.6.7 Serviceability Performance Functions – Overall and Inter-Story Drifts

As in the previous example, two required serviceability-related LSFs of overall and inter-story drifts are represented by Eq. 8.29. The permissible or allowable δ_{allow} values are summarized in Table 8.2 for IO, LS, and CP performance levels, as suggested in FEMA-350 (2000). Using the information given in Table 8.2, δ_{allow} values for the overall lateral drift for the three structures are considered to be 59.44, 185.85, and 403.65 cm, respectively, for the performance level CP. For the LS performance level, the corresponding values are 29.72 cm, 92.93 cm, and 201.83 cm, respectively. They are 8.32 cm, 26.02 cm, and 56.51 cm, respectively, for the IO performance level. For the inter-story drift LSF, and the performance levels of CP, LS, and IO, δ_{allow} values are 19.81 cm, 9.91 cm, and 2.77 cm on the 2nd, 4th, and 11th floors of the three buildings.

Table 8.9 Ground motions corresponding to 10% PE in 50 years and LS performance level, Site Class S_D.

EQ	Name	M_w	SF	PGA (g)	Time (sec)
21	Imperial Valley, 1940	6.9	2.01	0.461	25.0
22	Imperial Valley, 1940	6.9	2.01	0.675	25.0
23	Imperial Valley, 1979	6.5	1.01	0.393	15.0
24	Imperial Valley, 1979	6.5	1.01	0.488	15.0
25	Imperial Valley, 1979	6.5	0.84	0.301	15.0
26	Imperial Valley, 1979	6.5	0.84	0.234	15.0
27	Landers, 1992	7.3	3.20	0.421	30.0
28	Landers, 1992	7.3	3.20	0.425	30.0
29	Landers, 1992	7.3	2.17	0.519	30.0
30	Landers, 1992	7.3	2.17	0.360	30.0
31	Loma Prieta, 1989	7.0	1.79	0.665	16.0
32	Loma Prieta, 1989	7.0	1.79	0.969	16.0
33	Northridge, 1994, Newhall	6.7	1.03	0.678	15.0
34	Northridge, 1994, Newhall	6.7	1.03	0.657	15.0
35	Northridge, 1994, Rinaldi	6.7	0.79	0.533	14.0
36	Northridge, 1994, Rinaldi	6.7	0.79	0.579	14.0
37	Northridge, 1994, Sylmar	6.7	0.99	0.569	15.0
38	Northridge, 1994, Sylmar	6.7	0.99	0.817	15.0
39	North Palm Springs, 1986	6.0	2.97	1.018	16.0
40	North Palm Springs, 1986	6.0	2.97	0.986	16.0

8.7 Structural Reliability Evaluations of the Three Buildings for the Performance Levels of CP, LS, and IO Using REDSET

The uncertainty associated with all RVs present in the formulation, the earthquake time histories necessary to excite the three structures, and the LSFs are available at this stage. REDSET is used to estimate the reliability indexes β and corresponding p_f assuming all the connections are FR type or PR type with post-Northridge connections, and TDNFEA values for the two LSFs are summarized for the three buildings and for the three performance levels in Tables 8.11–8.16.

Table 8.10 Ground motions corresponding to 50% PE in 50 years and IO performance level, Site Class S_D.

EQ	Name	M_w	SF	PGA (g)	Time (sec)
41	Coyote Lake, 1979	5.7	2.28	0.589	12.0
42	Coyote Lake, 1979	5.7	2.28	0.333	12.0
43	Imperial Valley, 1979	6.5	0.40	0.143	15.0
44	Imperial Valley, 1979	6.5	0.40	0.112	15.0
45	Kern, 1952	7.7	2.92	0.144	30.0
46	Kern, 1952	7.7	2.92	0.159	30.0
47	Landers, 1992	7.3	2.63	0.337	25.0
48	Landers, 1992	7.3	2.63	0.307	25.0
49	Morgan Hill, 1984	6.2	2.35	0.318	20.0
50	Morgan Hill, 1984	6.2	2.35	0.546	20.0
51	Parkfield, 1966, Cholame	6.1	1.81	0.780	15.0
52	Parkfield, 1966, Cholame	6.1	1.81	0.631	15.0
53	Parkfield, 1966, Cholame	6.1	2.92	0.693	15.0
54	Parkfield, 1966, Cholame	6.1	2.92	0.790	15.0
55	North Palm Springs, 1986	6.0	2.75	0.517	20.0
56	North Palm Springs, 1986	6.0	2.75	0.379	20.0
57	San Fernando, 1971	6.5	1.30	0.253	20.0
58	San Fernando, 1971	6.5	1.30	0.231	20.0
59	Whittier, 1987	6.0	3.62	0.768	15.0
60	Whittier, 1987	6.0	3.62	0.478	15.0

8.7.1 Observations for the Three Performance Levels

Reliability information summarized in the six tables needs critical evaluation. The reliability index, β, values are very different for the three buildings when they are excited by 20 design earthquake time histories for the three performance levels. The results are expected since the frequency contents of each record are different. This observation justifies the use of multiple time histories to design a structure as suggested in more recent design guidelines (ASCE/SEI 7-10 2010). For all cases, the reliability information was extracted using only few hundreds of TNDFEA using REDSET. Furthermore, the reliability information is very similar using commonly used practices for connections of FR type and post-Northridge PR type connections. The results indicate that the

Table 8.11 Reliability corresponding to overall drift LSF and CP performance level, Site Class S_D.

	CP (2475-year return period) – overall drift					
	3-story		9-story		20-story	
	FR	PR	FR	PR	FR	PR
EQ	B (TNDFEA)	β (TNDFEA)	β (TNDFEA)	β (TNDFEA)	β (TNDFEA)	β (TNDFEA)
1	7.720 (209)	8.005 (239)	4.304 (341)	4.439 (341)	2.927 (379)	2.809 (394)
2	9.388 (209)	10.972 (209)	6.318 (341)	5.960 (341)	5.398 (379)	6.802 (379)
3	5.745 (209)	4.125 (209)	7.456 (356)	8.835 (341)	5.595 (394)	4.753 (379)
4	8.380 (209)	8.180 (209)	4.090 (356)	4.807 (341)	4.680 (379)	4.739 (379)
5	6.415 (209)	6.523 (209)	5.227 (356)	5.322 (341)	8.024 (394)	8.449 (379)
6	9.563 (209)	8.486 (209)	4.553 (356)	4.615 (341)	7.274 (379)	8.428 (379)
7	11.204 (209)	11.634 (209)	6.330 (341)	6.828 (341)	4.365 (379)	4.584 (409)
8	3.880 (209)	3.922 (209)	4.668 (356)	4.798 (356)	5.257 (379)	5.254 (379)
9	5.826 (209)	6.638 (209)	9.000 (341)	9.635 (356)	8.490 (394)	9.231 (379)
10	6.282 (209)	6.399 (239)	8.038 (341)	8.970 (341)	4.808 (409)	3.760 (409)
11	5.346 (209)	5.537 (209)	4.631 (341)	5.725 (341)	4.436 (379)	4.533 (379)
12	8.622 (209)	9.530 (209)	4.289 (341)	4.308 (341)	6.624 (379)	6.550 (409)
13	9.584 (209)	10.463 (209)	6.990 (341)	6.712 (341)	4.573 (379)	4.611 (379)
14	8.467 (209)	9.451 (209)	6.462 (341)	6.749 (341)	4.546 (409)	4.525 (409)
15	6.741 (209)	6.732 (209)	3.820 (341)	4.029 (341)	3.020 (394)	3.032 (409)
16	8.362 (209)	6.282 (209)	3.733 (341)	3.890 (356)	2.693 (379)	2.737 (409)
17	9.535 (209)	9.963 (209)	5.540 (356)	5.657 (341)	5.987 (379)	5.877 (379)
18	9.373 (224)	11.034 (209)	4.051 (356)	4.181 (341)	4.365 (409)	4.417 (379)
19	7.004 (209)	8.824 (209)	7.125 (341)	7.419 (356)	9.021 (379)	9.313 (409)
20	5.878 (224)	4.753 (224)	4.550 (356)	4.748 (341)	4.147 (379)	4.217 (379)

post-Northridge improvements are very beneficial and should be adopted as soon as practicable. In general, β values fall within a range, satisfying the intent of the code (ASCE/SEI 7-10 2010).

For ease of discussion, the mean values of β and the corresponding p_f for the three performance levels for the three buildings are given in Table 8.17.

The 3-story building designed by experts is not expected to develop any serviceability-related problem. It can be overlooked for additional discussions. Higher β values indicate the lower p_f values. One can suggest that p_f values should be higher

Table 8.12 Reliability corresponding to inter-story drift LSF and CP performance level, Site Class S_D.

	CP (2475-year return period) – inter-story drift					
	3-story		9-story		20-story	
	FR	PR	FR	PR	FR	PR
EQ	β (TNDFEA)	β (TNDFEA)	β (TNDFEA)	β (TNDFEA)	β (TNDFEA)	β (TNDFEA)
1	8.393 (239)	8.761 (209)	3.830 (341)	3.952 (341)	3.849 (379)	3.887 (409)
2	9.512 (209)	9.790 (209)	5.999 (341)	5.553 (356)	5.500 (409)	5.242 (379)
3	5.925 (209)	5.374 (239)	6.206 (341)	7.062 (341)	5.075 (379)	5.186 (379)
4	8.571 (209)	8.524 (209)	4.690 (356)	4.421 (371)	4.292 (379)	4.388 (379)
5	5.836 (209)	5.896 (209)	4.723 (341)	4.796 (341)	5.701 (394)	5.720 (394)
6	7.681 (209)	8.169 (239)	4.175 (356)	4.219 (356)	3.835 (409)	4.015 (379)
7	8.709 (209)	9.021 (209)	5.969 (341)	6.342 (341)	4.581 (379)	4.584 (409)
8	5.520 (209)	5.475 (209)	4.055 (341)	4.169 (356)	4.412 (394)	4.436 (379)
9	4.962 (209)	5.480 (239)	8.296 (356)	8.564 (341)	7.474 (379)	7.469 (379)
10	5.814 (209)	5.972 (209)	7.587 (341)	8.157 (356)	3.230 (379)	3.620 (409)
11	4.742 (209)	5.010 (209)	4.135 (341)	4.255 (341)	4.549 (379)	5.984 (409)
12	7.060 (239)	7.721 (239)	3.938 (356)	3.920 (371)	3.504 (409)	3.795 (379)
13	9.209 (209)	8.598 (209)	6.939 (341)	7.332 (341)	4.116 (379)	4.170 (379)
14	7.805 (209)	8.643 (209)	4.419 (341)	4.589 (341)	4.287 (379)	4.260 (409)
15	5.478 (209)	5.818 (209)	3.468 (356)	3.672 (356)	2.337 (379)	2.429 (379)
16	8.748 (239)	8.778 (209)	3.334 (341)	3.498 (341)	2.321 (379)	2.435 (409)
17	7.385 (209)	7.279 (209)	5.300 (341)	5.385 (356)	6.773 (394)	5.993 (409)
18	8.782 (239)	9.842 (209)	3.643 (356)	3.775 (341)	4.048 (379)	4.127 (409)
19	7.266 (209)	7.714 (209)	6.643 (341)	7.013 (341)	8.119 (409)	7.968 (379)
20	6.611 (239)	6.093 (209)	4.143 (356)	4.345 (341)	3.922 (394)	3.997 (409)

when excited by the most severe ground motion thinking in a deterministic way. The results in Table 8.17 indicate that for each performance level and considering 9- and 20-story structures, the average reliability indexes, β_{average} values are within a narrow range, confirming the intent of the PBSD guidelines (FEMA-350 2000). The severity levels of earthquakes and the corresponding δ_{allow} values are taken into account in defining the LSFs and similarities in the β values are expected. This observation is very encouraging for the implementation of PBSD using REDSET. In fact, REDSET demonstrates a very highly desirable feature for any acceptable

Table 8.13 Reliability corresponding to overall drift LSF and LS performance level, Site Class S_D.

	LS (475-year return period) – overall drift					
	3-story		9-story		20-story	
	FR	PR	FR	PR	FR	PR
EQ	β (TNDFEA)	β (TNDFEA)	β (TNDFEA)	β (TNDFEA)	β (TNDFEA)	β (TNDFEA)
21	4.929 (209)	5.605 (209)	6.690 (341)	6.969 (341)	4.549 (379)	4.910 (409)
22	6.804 (209)	7.172 (209)	6.701 (341)	6.754 (341)	6.302 (409)	6.417 (379)
23	4.180 (224)	4.332 (209)	6.795 (341)	7.017 (356)	4.199 (379)	5.708 (379)
24	7.636 (209)	7.321 (209)	2.737 (356)	3.916 (341)	8.169 (409)	8.381 (409)
25	7.594 (209)	7.117 (209)	7.369 (341)	7.487 (341)	6.549 (409)	6.585 (379)
26	6.115 (209)	6.020 (209)	6.854 (356)	6.777 (341)	6.806 (379)	6.258 (394)
27	6.725 (224)	6.720 (209)	7.168 (341)	7.177 (371)	3.183 (394)	3.175 (409)
28	9.284 (209)	11.395 (209)	6.778 (356)	6.832 (341)	4.753 (409)	4.016 (394)
29	7.774 (224)	7.443 (209)	3.473 (356)	3.634 (341)	6.580 (379)	6.083 (379)
30	10.246 (209)	9.776 (209)	5.416 (341)	5.468 (356)	4.949 (379)	4.021 (379)
31	5.638 (209)	5.729 (209)	4.309 (341)	4.429 (341)	4.309 (379)	4.320 (379)
32	4.061 (209)	4.189 (224)	8.302 (356)	8.551 (371)	3.378 (379)	3.832 (379)
33	4.583 (209)	4.900 (209)	5.102 (341)	4.967 (356)	5.578 (379)	5.851 (379)
34	4.736 (209)	5.820 (209)	3.933 (356)	4.005 (341)	5.651 (409)	5.479 (409)
35	5.520 (209)	5.613 (209)	4.310 (356)	4.368 (341)	6.837 (409)	6.750 (379)
36	8.521 (209)	8.865 (209)	3.766 (341)	3.828 (356)	4.271 (379)	4.402 (409)
37	3.720 (209)	4.771 (209)	5.446 (356)	5.767 (356)	4.235 (379)	4.299 (379)
38	3.103 (209)	3.144 (209)	3.904 (356)	4.032 (356)	4.396 (409)	4.393 (379)
39	4.164 (224)	3.915 (209)	6.854 (341)	6.932 (356)	4.133 (409)	4.753 (409)
40	2.727 (209)	4.660 (209)	4.048 (341)	4.120 (356)	4.862 (379)	4.860 (379)

reliability evaluation method. For most of the cases, β_{average} values for the overall lateral drift are found to be higher than the inter-story drift, indicating the latter is more critical. The same behavior is observed in other examples considered in other chapters. It is also interesting to note the distinct separations of β values in Table 8.17 for the performance levels of CP, LS, and IO. This indicates that the SFs suggested by Somerville (1997) to match TPGRS at certain frequencies are rational. The results also demonstrate that the performances of buildings are very similar when connections are considered to be FR and post-Northridge PR types.

Table 8.14 Reliability corresponding to inter-story drift LSF and LS performance level, Site Class S_D.

	LS (475-year return period) – inter-story drift					
	3-story		9-story		20-story	
	FR	PR	FR	PR	FR	PR
EQ	β (TNDFEA)	β (TNDFEA)	β (TNDFEA)	β (TNDFEA)	β (TNDFEA)	β (TNDFEA)
21	9.156 (224)	9.431 (209)	4.295 (356)	4.658 (341)	4.450 (409)	4.854 (409)
22	4.777 (209)	4.643 (209)	6.426 (341)	6.430 (356)	4.600 (379)	4.728 (379)
23	8.105 (209)	8.295 (209)	6.333 (341)	6.575 (356)	5.708 (409)	5.850 (409)
24	6.789 (209)	6.732 (209)	10.079 (341)	9.086 (356)	6.304 (379)	7.274 (409)
25	10.083 (224)	10.668 (209)	6.871 (356)	7.036 (341)	4.632 (409)	4.961 (379)
26	7.976 (224)	8.156 (209)	6.988 (341)	7.116 (356)	7.578 (379)	8.245 (409)
27	8.505 (224)	8.540 (209)	6.645 (371)	6.663 (356)	6.481 (409)	6.164 (394)
28	9.884 (224)	10.199 (209)	6.367 (341)	6.432 (356)	7.003 (379)	7.012 (409)
29	9.072 (209)	9.608 (209)	3.153 (356)	3.315 (356)	3.977 (409)	3.028 (409)
30	9.811 (209)	8.912 (209)	5.098 (341)	5.036 (356)	3.556 (379)	3.111 (379)
31	8.491 (209)	8.455 (224)	3.935 (356)	4.078 (356)	3.603 (379)	3.671 (394)
32	6.047 (209)	6.128 (224)	7.973 (371)	8.858 (356)	3.010 (409)	2.847 (379)
33	4.126 (209)	4.457 (209)	4.378 (341)	4.443 (356)	3.257 (379)	3.479 (409)
34	8.141 (209)	8.706 (224)	3.466 (356)	3.536 (341)	4.058 (409)	4.033 (409)
35	4.960 (224)	5.023 (209)	3.933 (371)	4.008 (356)	4.764 (379)	4.769 (409)
36	8.898 (209)	9.374 (209)	3.392 (356)	3.439 (356)	3.190 (379)	3.337 (379)
37	9.661 (209)	10.774 (209)	5.111 (356)	5.353 (341)	3.790 (394)	3.780 (409)
38	4.559 (209)	4.612 (209)	3.292 (356)	3.410 (356)	3.604 (379)	3.627 (379)
39	5.738 (224)	5.724 (209)	6.640 (341)	6.846 (341)	2.860 (379)	3.793 (409)
40	5.592 (209)	6.511 (209)	3.595 (341)	3.628 (341)	3.732 (409)	3.786 (379)

The presence of two slots in the web of the beam in beam-column connections does not reduce the stiffness of the post-Northridge PR connections; in fact, they improve their behavior. The study also indicates that post-Northridge connections significantly improve the behavior of structures, and the rigidities of connections need to be appropriately incorporated into any risk-evaluation algorithm. For seismically active regions, designing beam-to-column connections as post-Northridge PR type instead of FR type is expected to be very economical since FR connections are very expensive to build.

Table 8.15 Reliability corresponding to overall drift LSF and IO performance level, Site Class S_D.

	IO (72-year return period) – overall drift					
	3-story		9-story		20-story	
	FR	PR	FR	PR	FR	PR
EQ	β (TNDFEA)	β (TNDFEA)	β (TNDFEA)	β (TNDFEA)	β (TNDFEA)	β (TNDFEA)
41	2.145 (224)	2.230 (209)	2.682 (356)	2.722 (356)	2.771 (379)	2.770 (409)
42	6.510 (209)	6.483 (224)	3.623 (356)	3.773 (341)	4.215 (394)	4.214 (379)
43	8.993 (209)	9.184 (209)	4.560 (356)	4.676 (356)	3.344 (379)	3.508 (394)
44	10.014 (224)	10.537 (209)	7.412 (341)	7.414 (341)	2.020 (379)	2.452 (409)
45	8.020 (209)	8.304 (209)	4.880 (356)	4.954 (341)	4.323 (409)	4.411 (379)
46	8.685 (209)	8.895 (224)	5.111 (341)	5.120 (341)	4.484 (379)	4.299 (409)
47	4.217 (209)	4.393 (209)	3.595 (341)	3.814 (356)	2.493 (394)	2.164 (394)
48	6.404 (209)	6.981 (209)	7.125 (356)	7.390 (341)	2.372 (379)	2.010 (409)
49	6.905 (209)	6.207 (224)	3.638 (341)	3.850 (356)	2.599 (409)	2.580 (379)
50	5.105 (224)	5.088 (224)	4.502 (341)	5.205 (341)	4.475 (409)	4.676 (379)
51	2.162 (209)	2.174 (224)	4.325 (356)	4.416 (371)	3.138 (379)	3.070 (409)
52	2.140 (209)	2.219 (224)	7.049 (341)	7.016 (356)	7.675 (379)	7.969 (379)
53	4.853 (209)	4.922 (209)	4.238 (356)	4.274 (356)	6.686 (394)	7.121 (409)
54	3.297 (224)	3.368 (209)	3.980 (356)	4.465 (356)	4.083 (409)	4.038 (379)
55	5.681 (209)	5.994 (209)	3.124 (341)	3.191 (356)	4.405 (379)	4.346 (409)
56	7.650 (209)	7.864 (224)	3.145 (356)	3.121 (341)	3.325 (409)	3.001 (379)
57	7.076 (224)	7.845 (224)	5.852 (341)	5.879 (341)	4.505 (379)	4.187 (394)
58	7.989 (209)	7.930 (209)	3.345 (341)	3.496 (356)	3.287 (409)	3.362 (379)
59	2.284 (209)	2.603 (224)	1.447 (341)	1.495 (356)	2.998 (379)	3.096 (379)
60	2.404 (224)	2.601 (209)	1.449 (341)	1.362 (341)	4.385 (409)	4.379 (379)

As mentioned earlier, the patented post-Northridge PR connections with slotted holes in the web do not compromise the rigidities of the connections, as observed in full-scale testing. To analytically document this observation, reliability index β values for FR and post-Northridge PR types for 3-, 9- and 20-story buildings are plotted in Figures 8.12–8.14, respectively. In these figures, a linear trend along a 45° indicates that their behavior is very similar as observed during experimental investigations. However, a higher spread can be observed for the overall lateral drift in comparison with the inter-story drift.

Table 8.16 Reliability corresponding to inter-story drift LSF and IO performance level, Site Class S_D.

	IO (72-year return period) – inter-story drift					
	3-story		9-story		20-story	
	FR	PR	FR	PR	FR	PR
EQ	β (TNDFEA)	β (TNDFEA)	β (TNDFEA)	β (TNDFEA)	β (TNDFEA)	β (TNDFEA)
41	1.548 (209)	1.644(209)	2.277 (341)	2.194 (341)	2.575 (394)	2.751 (409)
42	5.873 (209)	5.851 (224)	3.119 (371)	3.302 (371)	3.598 (379)	3.670 (379)
43	9.621 (224)	9.964 (224)	4.287 (356)	4.368 (341)	3.026 (379)	3.214 (379)
44	9.142 (209)	9.762 (224)	6.890 (356)	6.919 (356)	4.500 (379)	4.811 (409)
45	7.423 (209)	7.555 (224)	4.352 (341)	4.503 (356)	3.525 (409)	3.854 (379)
46	7.863 (224)	8.025 (224)	4.610 (356)	4.632 (341)	4.071 (394)	4.140 (409)
47	3.704 (209)	3.885 (209)	3.202 (371)	3.416 (341)	2.419 (409)	2.171 (379)
48	5.559 (224)	6.000 (209)	6.671 (341)	7.554 (356)	2.070 (379)	2.000 (409)
49	4.584 (209)	4.686 (209)	3.192 (341)	3.414 (356)	2.120 (409)	2.064 (379)
50	4.529 (209)	4.504 (224)	4.504 (356)	4.390 (341)	3.142 (379)	4.483 (394)
51	1.568 (224)	1.584 (209)	4.101 (356)	4.146 (371)	2.195 (409)	2.416 (409)
52	1.561 (209)	1.652 (209)	6.201 (341)	7.502 (356)	3.907 (379)	2.736 (394)
53	4.266 (209)	4.290 (224)	3.666 (356)	3.791 (356)	6.125 (394)	6.405 (409)
54	2.766 (224)	2.843 (209)	3.860 (341)	4.396 (356)	2.436 (409)	2.614 (379)
55	5.086 (209)	5.331 (209)	2.730 (356)	2.775 (341)	2.964 (379)	3.085 (409)
56	5.611 (209)	5.918 (224)	2.695 (341)	2.691 (356)	4.913 (409)	4.020 (394)
57	6.362 (224)	6.947 (209)	5.756 (356)	5.653 (341)	2.676 (409)	2.621 (379)
58	7.598 (209)	7.276 (224)	2.996 (341)	3.139 (356)	3.809 (379)	2.743 (409)
59	2.347 (224)	2.433 (224)	1.406 (356)	1.439 (341)	3.500 (409)	3.450 (379)
60	1.892 (224)	2.104 (209)	2.017 (341)	1.946 (356)	2.362 (379)	2.226 (394)

The exercise conclusively confirmed the response behavior of buildings is significantly improved in the presence of post-Northridge PR connections when excited by the earthquake time histories. Results in the three figures clearly demonstrate that post-Northridge beam-to-column connections are beneficial and demonstrate that the profession is moving in the right direction in designing more seismic risk-tolerant and resilient steel structures.

With the help of several numerical examples, the accuracy, efficiency, and robustness REDSET in implementing PBSD are demonstrated and documented

Table 8.17 Mean Structural reliability ($\beta\mu$) of 3-, 9-, and 20-story buildings, Site Category S_D.

Performance level	Structure	Overall lateral drift		Inter-story drift	
		FR	PR	FR	PR
		$\beta_{average}$ (p_f)	$\beta_{average}$ (p_f)	$\beta_{average}$ (p_f)	$\beta_{average}$ (p_f)
CP (2475-year return period)	3-story	7.666 (8.8818E−15)	7.873 (1.7764E−15)	7.200 (3.0109E−13)	7.398 (6.9167E−14)
	9-story	5.559 (1.3566E−08)	5.881 (2.0390E−09)	5.075 (1.9375E−07)	5.251 (7.5638E−08)
	20-story	5.311 (5.4513E−08)	5.431 (2.8020E−08)	4.596 (2.1534E−06)	4.685 (1.3998E−06)
LS (475-year return period)	3-story	5.903 (1.7848E−09)	6.225 (2.4078E−10)	7.518 (2.7756E−14)	7.747 (4.6629E−15)
	9-story	5.498 (1.9206E−08)	5.652 (7.9296E−09)	5.399 (3.3507E−08)	5.497 (1.9315E−08)
	20-story	5.184 (1.0859E−07)	5.225 (8.7078E−08)	4.508 (3.2721E−06)	4.617 (1.9466E−06)
IO (72-year return period)	3-story	5.627 (9.1685E−09)	5.791 (3.4984E−09)	4.945 (3.8072E−07)	5.113 (1.5854E−07)
	9-story	4.254 (1.0499E−05)	4.382 (5.8797E−06)	3.927 (4.3006E−05)	4.109 (1.9869E−05)
	20-story	3.879 (5.2443E−05)	3.883 (5.1588E−05)	3.297 (4.8862E−04)	3.274 (5.3018E−04)

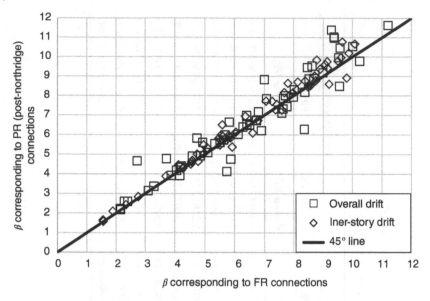

Figure 8.12   Reliability comparison among FR and PR (post-Northridge) connections for a 3-story building.

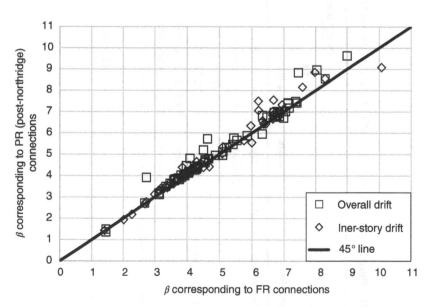

Figure 8.13   Reliability comparison among FR and PR (post-Northridge) connections for a 9-story building.

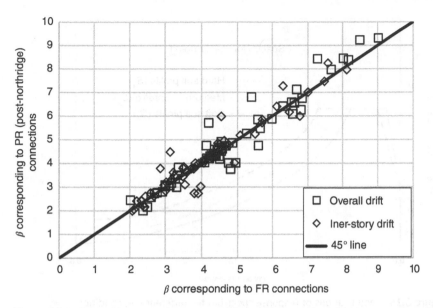

Figure 8.14 Reliability comparison among FR and PR (post-Northridge) connections for a 20-story building.

in this section. Engineers interested in using the PBSD concept in the seismically active region can follow the outline of the procedure presented here. It will provide economical, dependable, and more seismic load-tolerant structures.

8.8 Implementation of PBSD for Different Soil Conditions

After detailed post-earthquake damage assessments, numerous changes are generally proposed in building codes to avoid similar occurrences. For example, the Mexico City earthquake of magnitude 8.1 on 19 September 1985 caused significant damage to infrastructures. Mexico City is located on a dry lakebed with poor soil conditions. The epicentral distance of the earthquake was about 400 km. The extensive damages caused by the earthquake were attributed to the poor soil conditions in the region. The incidence caused major changes in seismic design guidelines. To address poor soil conditions, generally the peak design response spectrum is extended over larger periods. The modification is conceptually shown in Figure 8.15.

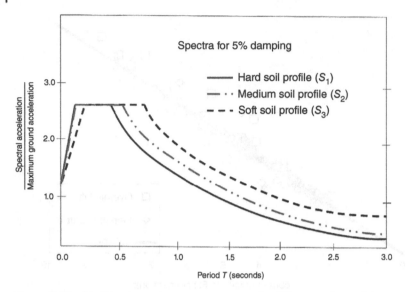

Figure 8.15 Modifications of response spectrum for different soil conditions.

Unfortunately, again on 19 September 2017, another earthquake of magnitude 7.1 occurred in Mexico City. This time the epicentral distance was 120 km. It also caused a lot of damages. It is difficult to determine whether the damaged structures were built before or after 1985. In any case, historical records also indicate that all structures in a region with poor soil conditions do not suffer similar damages; in fact, some of them remain damage-free. This observation indicates that the underlying dynamics under different soil conditions caused by seismic excitation need to be addressed more appropriately.

There are many reasons for undesirable behavior of infrastructures located in poor soil conditions including difficulty in accurately mathematically modeling the complicated soil-structure systems, predicting the future design earthquake time history at the site considering the soil condition, modeling and evaluating the dynamic amplification of responses caused by the excitation, incorporating major sources of nonlinearity and energy dissipation in the formulation, and most importantly considering the presence of a considerable amount of uncertainties at every phase of the evaluation process. Since soil is not man-made, a considerable amount of non-homogeneity and uncertainty is expected in modeling it. When an earthquake occurs, seismic waves travel from the hypocenter to a specific site through various soil conditions. Because of this, it is extremely difficult to predict the frequency contents of the design earthquake at a site. A structure resting on a site will have its own specific dynamic characteristics and frequency contents.

Obviously, if the structural and the exciting earthquake frequencies approach the resonance condition, the structure is expected to suffer a significant amount of damages. In the United States, to address the unpredictable frequency contents of the design earthquake, ASCE 7-10 (2010) requires that a structure needs to be excited by multiple similar design earthquake time histories fitting a site-specific TPGRS, initially at least 7 and now at least 11 (Zimmerman et al. 2015). The ASCE guidelines did not explicitly address modeling uncertainties in structural parameters, LL (assuming DL can be accurately estimated with a small amount of uncertainty), and several sources of energy dissipations. To satisfy the intents of the code, uncertainty associated with all the design variables in addition to the earthquake excitation, needs to be considered appropriately. The basic idea is that the underlying risk needs to be realistically estimated by conducting multiple deterministic analyses of a structure excited by several site-specific design earthquake time histories in the presence of several major sources of uncertainty. If developed appropriately, this concept may require few dozen instead of 7 or 11 deterministic analyses, as suggested in the current design guidelines. The RED-SET approach will be ideal to estimate the underlying seismic risk with different soil conditions at a site satisfying the basic intent of the code.

For the sake of completeness, soil types are classified into 6 categories in the United States (ASCE 7-10 2010). They are denoted as Site Classes A–F. Site Class A represents hard rock with measured shear wave velocity greater than 1500 m/s. Site Class F requires site-specific evaluations. FEMA guidelines also suggest that when the soil properties are not known in sufficient detail, Site Class D shall be used. To consider different soil conditions, for the illustrative examples given next, a structure is considered to be located on two types of soil, Site Classes B and D.

8.9 Illustrative Example of Reliability Estimation for Different Soil Conditions

A 9-story steel-frame building, discussed in Section 8.6 and shown in Figure 8.10, is considered to showcase the risk estimation potential of REDSET to implement PBSD with different soil conditions; Site Class B – representing rock with shear wave velocity between 760 m/s and 1500 m/s and Site Class D – no information is available on the soil condition at the site. As discussed earlier, SAC designed the building satisfying the 1997 NEHRP provisions for a site in Los Angeles with four different column sizes (W14, W24, W30, and W36) (FEMA-355F 2000). Buildings designed using W14 columns with a fundamental period of 2.18 seconds are considered for this example.

8.9.1 Quantifications of Uncertainties for Resistance-Related Variables and Gravity Loads

Uncertainties associated with all the resistance and gravity load-related variables listed in Table 8.7 will remain the same.

8.9.2 Generation of Multiple Design Earthquake Time Histories for Different Soil Conditions

Besides using time histories suggested by (Somerville 1997), other alternatives discussed in Chapter 4 can also be used. They are Alternative 1 – scaling earthquake time histories available at PEER database and Alternative 2 – using detailed information on seismic activities of the region using BBP. Alternative 1 is used in this section. This approach is expected to be suitable for worldwide applications.

For this study, the ground motion time histories were selected by matching the Uniform Hazard Spectrum (UHS) for performance levels CP, LS, and IO. For a site in the Los Angeles region, TPGRS and UHS will be the same. Performance levels CP, LS, and IO correspond to a 2%, 10%, and 50% PE in 50 years, respectively. Following the procedure discussed in Chapter 4 for Alternative 1, a total of six suites (for 2 soil types and 3 performance levels), each containing 11 most suitable scaled ground motions, are selected. They are listed in Tables 8.18 and 8.19 for soil types B and D, respectively, giving information on the name of the original earthquake and the SF used.

Spectral accelerations for 11 earthquakes for Site Class D and CP performance level are plotted in Figure 8.16 indicating that the ground motions in the database are scaled to match the TPGRS at the fundamental period of the structure; for this example, it is 2.18 seconds. Similar plots can be made for Site Class B and D and other performance levels.

8.9.3 Implementation of PBSD for Different Soil Conditions

Information on all major sources of uncertainty and multiple acceleration time histories of earthquake loading for the three performance functions and two soil conditions is now available. The required IRSs for each performance level and soil type are generated by conducting TNDFEA of about 300 deterministic analyses by exciting the frame by the corresponding earthquake time histories. With the availability of IRSs and the allowable values discussed in detail for the 9-story building in Section 8.6.1, the required LSFs of overall and inter-story drift at the 5th-floor level are now available. The reliability indexes β values are estimated using REDSET for the two LSFs for Soil Classes B and D and the information is summarized in Tables 8.20 and 8.21, respectively.

Table 8.18 Sets of ground motions for soil type B – CP, LS, and IO performance levels.

Set 1: 2% PE in 50 years; CP			Set 2: 10% PE in 50 years; LS			Set 3: 50% PE in 50 years; IO		
EQ	Record name	SF	EQ	Record name	SF	EQ	Record name	SF
1	1971 San Fernando	0.48	12	1971 San Fernando	0.26	23	1971 San Fernando	0.10
2	1995 Kobe, Japan	0.60	13	1989 Loma Prieta	0.38	24	1989 Loma Prieta	0.43
3	1989 Loma Prieta	0.71	14	1980 Iripinia, Italy	0.53	25	1994 Northridge-01	0.55
4	1999 Chi-Chi, Taiwan	1.33	15	1999 Kocaeli, Turkey	0.94	26	1989 Loma Prieta	0.60
5	1980 Irpinia, Italy-01	1.70	16	1989 Loma Prieta	1.15	27	1989 Loma Prieta	0.65
6	1999 Kocaeli, Turkey	1.77	17	1989 Loma Prieta	1.73	28	1989 Loma Prieta	1.18
7	1980 Irpinia, Italy-02	2.22	18	1989 Loma Prieta	2.34	29	1989 Loma Prieta	1.24
8	1999 Kocaeli, Turkey	2.36	19	1994 Northridge-01	2.56	30	1980 Iripinia, Italy	1.41
9	1994 Northridge-01	2.77	20	1989 Loma Prieta	3.16	31	1971 San Fernando	1.93
10	1994 Northridge-01	2.89	21	1989 Loma Prieta	3.33	32	1984 Morgan Hill	3.25
11	1989 Loma Prieta	3.24	22	1994 Northridge-01	3.98	33	1984 Morgan Hill	3.85

Several important observations can be made from the results given in Tables 8.20 and 8.21. Although the earthquake time histories were selected very carefully satisfying the TPGRS for soil class and the structure, the β values are quite different. The authors considered other similar buildings suggested by FEMA (FEMA-355F 2000) and observed a similar spread in the β values. Considering that any one of the design earthquake time histories can occur at the site, it is obvious that the absolute safety of the structure cannot be assured. However, considering 11 design time histories, the structure can be designed as more damage tolerant. This observation clearly indicates that the frequency contents of an earthquake are a very

Table 8.19 Sets of ground motions for soil type D for CP, LS, and IO performance levels.

Set 4: 2% PE in 50 years; CP			Set 5: 10% PE in 50 years; LS			Set 6: 50% PE in 50 years; IO		
EQ	Record name	SF	EQ	Record name	SF	EQ	Record name	SF
34	1995 Kobe, Japan	0.63	45	1995 Kobe, Japan	0.34	56	1994 Northridge-01	0.17
35	1994 Northridge-01	1.59	46	1994 Northridge-01	0.87	57	1995 Kobe, Japan	0.43
36	1979 Imperial Valley-06	2.24	47	1995 Kobe, Japan	1.30	58	1995 Kobe, Japan	0.50
37	1989 Loma Prieta	2.37	48	1995 Kobe, Japan	2.01	59	1995 Kobe, Japan	0.81
38	1995 Kobe, Japan	2.40	49	1995 Kobe, Japan	2.12	60	1995 Kobe, Japan	0.84
39	1999 Chi-Chi, Taiwan	2.71	50	1999 Chi-Chi, Taiwan	2.13	61	1995 Kobe, Japan	0.93
40	1979 Imperial Valley-06	3.13	51	1995 Kobe, Japan	2.18	62	1983 Coalinga-05	1.12
41	1999 Chi-Chi, Taiwan	3.48	52	1995 Kobe, Japan	2.41	63	1995 Kobe, Japan	1.48
42	1995 Kobe, Japan	3.70	53	1983 Coalinga-05	2.92	64	1999 Chi-Chi, Taiwan	1.56
43	1999 Chi-Chi, Taiwan	3.85	54	1983 Coalinga-01	3.23	65	1983 Coalinga-05	1.68
44	1995 Kobe, Japan	3.90	55	1995 Kobe, Japan	3.84	66	1966 Parkfield	3.24

critical design parameter but they are beyond the control of designers. The reliability indexes did not change considerably for CP, LS, and IO, as expected for a particular time history, in most cases. This is expected since allowable values will be different for the performance level. Similar behavior was also observed in the previous examples. This observation clearly indicates that the selection process of the appropriate time histories is essential and should not be overlooked in routine applications. In most cases, the building located on a site with Soil Class B has a higher β value than that of Soil Class D, but not always. Inter-story drift appears to be more critical than the overall drift, as observed in other examples. It also

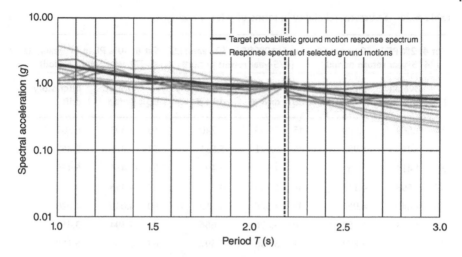

Figure 8.16 Spectral accelerations for 11 earthquakes for Site Class D and CP performance level.

Table 8.20 Structural reliability in terms of β for B soil type.

	Set 1: 2% PE in 50 years; CP (2475-year return period)			Set 2: 10% PE in 50 years; LS (475-year return period)			Set 3: 50% PE in 50 years; IO (72-year return period)	
EQ	Overall drift β	Inter-story drift β	EQ	Overall drift β	Inter-story drift β	EQ	Overall drift β	Inter-story drift β
1	8.533	4.753	12	8.072	6.245	23	9.333	10.343
2	7.828	6.977	13	9.625	6.059	24	6.011	3.230
3	2.548	3.618	14	9.056	7.097	25	6.200	10.204
4	1.651	3.387	15	7.476	6.792	26	11.679	11.517
5	4.776	1.166	16	5.558	3.558	27	4.583	6.536
6	4.753	4.753	17	4.585	7.170	28	7.772	10.807
7	5.556	3.582	18	5.174	6.378	29	5.942	7.370
8	7.316	7.917	19	9.365	7.228	30	4.434	6.469
9	5.405	4.753	20	3.380	2.195	31	10.772	9.890
10	9.918	9.481	21	5.520	2.248	32	10.129	6.341
11	5.992	5.479	22	11.104	6.286	33	3.115	5.331

Table 8.21 Structural reliability in terms of β for D soil type.

	Set 4: 2% PE in 50 years; CP (2475-year return period)		Set 5: 10% PE in 50 years; LS (475-year return period)			Set 6: 50% PE in 50 years; IO (72-year return period)		
EQ	Overall drift β	Inter-story drift β	EQ	Overall drift β	Inter-story drift β	EQ	Overall drift β	Inter-story drift β
34	3.869	4.980	45	5.895	5.740	56	6.393	6.502
35	6.448	5.982	46	6.021	5.581	57	5.375	5.363
36	4.414	5.851	47	4.959	5.192	58	5.455	5.455
37	5.598	4.115	48	5.134	4.715	59	4.446	4.259
38	4.640	4.955	49	4.328	4.212	60	4.046	4.020
39	4.753	5.782	50	3.696	3.666	61	4.394	3.452
40	4.231	3.873	51	5.396	5.392	62	5.335	5.168
41	7.888	4.121	52	4.240	3.061	63	4.311	4.219
42	4.377	4.664	53	6.733	6.553	64	6.105	6.204
43	4.097	4.068	54	3.539	3.719	65	7.279	7.182
44	4.132	4.177	55	5.635	5.541	66	3.702	3.478

documents the necessity of considering realistic physics-based dynamic soil conditions and structural characteristics to make an appropriate design decision. More emphasis should be given to the selection of the appropriate performance criterion. By conducting few dozens of deterministic analyses at very intelligently selected sampling points, hopefully using Sampling Scheme MS3-KM, structures can be designed more seismic load-tolerant. The example showcases how PBSD can be implemented, and the underlying risk can be estimated using REDSET. The concept is expected to change the current engineering design paradigm.

8.10 Concluding Remarks

Different capabilities of REDSET in estimating the risk of CNDES are demonstrated in this chapter. The underlying risk depends on the accuracy of the FE representation of the structure. Joints and support conditions need to be represented by the appropriate physics-based formulation. The flexibility of joints is now being used to improve the structural response behavior. It is a step in the right direction. It is also economical since PR connections are more economical to build. The recently introduced PBSD concept appears to be appealing. It is expected to be economical and helps design more seismic load-tolerant structures. The robustness of REDSET is demonstrated in this chapter by representing structures using the assumed stress-based FEM representation instead of displacement-based FEM similar to the process used in the SFEM concept. As mentioned earlier, REDSET will replace SFEM. Soil condition at a site is extremely important in selecting the appropriate time histories of the design earthquake. In most cases, the spread in the β values for different soil conditions is not small indicating the presence of a considerable amount of uncertainty. Since soil is not man-made, it is very non-homogeneous. An appropriate representation of soil conditions at a site is expected to remain a challenging problem in future designs. However, reducing the underlying risk using REDSET will be beneficial and will help design more seismic-risk tolerant structures making them more sustainable and resilient.

9

Reliability Assessment of Lead-Free Solders in Electronic Packaging Using REDSET for Thermomechanical Loadings

9.1 Introductory Comments

In the previous chapters, the underlying risk of onshore and offshore structures excited by the earthquakes and wave loadings in the time domain was estimated using REDSET. The implementation potential of the novel risk assessment procedure was also verified using different computational platforms. The accuracy, efficiency, and robustness of the procedure are now well established. It is now necessary to document that the proposed procedure is very advanced and can be used to estimate the underlying risk of different types of engineering structures excited by the thermomechanical loading in the time domain. To demonstrate the advanced features and robustness of REDSET, the underlying risk of lead-free solders used in electronic packaging (EP) subjected to thermomechanical loading caused by heating and cooling is presented in this chapter. The quantifications of uncertainty in the system's parameters and the information on modeling of the thermomechanical loading were not readily available in the literature. When a paper on the topic was submitted for publication in the Journal of Electronic Packaging published by the American Society of Mechanical Engineering, one reviewer thought the paper was groundbreaking (Azizsoltani and Haldar 2018). The major objective of this chapter is to encourage exploration of reliability estimation of other uncommon engineering systems with very little information on uncertainty in the design parameters and excitement by a rare form of dynamic loading using REDSET.

Solder balls used in EP are subjected to thermomechanical loading caused by the heating and cooling of the computer. It is a different type of dynamic loading, very different from earthquake and wave loadings. Conceptually, thermomechanical loading is also a time domain dynamic excitation problem in the presence of many sources of nonlinearity including severe material nonlinearity. Heating and cooling cause significant changes in the material properties of solder balls.

Reliability Evaluation of Dynamic Systems Excited in Time Domain: Alternative to Random Vibration and Simulation, First Edition. Achintya Haldar, Hamoon Azizsoltani, J. Ramon Gaxiola-Camacho, Sayyed Mohsen Vazirizade, and Jungwon Huh.

In the absence of any analytical study, solder balls were tested in laboratories to study their behavior (Whitenack 2004; Sane 2007). Laboratory tests are conducted under specific conditions of samples and a set of design variables. The outcomes of laboratory testing will be valid for the test conditions and samples used in the testing. It may not be possible to extrapolate results with slight variations of samples used or test conditions. To study the uncertainty-related issues, the testing program needs to be expanded causing additional expenses and time. Since the duration of testing can be very long spanning over several years, accelerated testing conditions are used to replicate the operating conditions introducing another major source of uncertainty. The reliability of solder balls cannot be estimated using any analytical methods currently available. Instead of conducting laboratory investigations, the analytical approach is a viable alternative to extract necessary reliability information when such information is not readily available. It will also help document changes in the behavior if the properties of the material and testing conditions are changed without conducting any additional tests saving a considerable amount of time and money. In general, laboratory experiments can be very expensive, and the collected data will have very limited use, applicable only to the test specimens and laboratory conditions used during testing. Moreover, the test results are proprietary in nature and may not be widely available to other scholars interested in the experiment. Instead of conducting laboratory investigations, an analytical approach is a viable approach to generate necessary reliability information and provides considerable flexibility to consider issues not considered during testing.

Analytical assessment is expected to be complicated and generally avoided since research activities in the related areas are limited. The limited literature on the topic attests to the statement. This chapter will demonstrate that the underlying risk can still be estimated even when available information on uncertainty is not well established. This exercise will extend the application boundaries of REDSET and showcase how the reliability information of other similar but complicated engineering systems can be extracted using it.

A solder ball is represented by finite elements. Major sources of nonlinearities are incorporated as realistically as practicable. Uncertainties in all design variables are quantified using available information. The thermomechanical loading is represented by five design parameters and uncertainties associated with them are incorporated. The accuracy, efficiency, and application potential of REDSET are established with the help of Monte Carlo simulation (MCS) and results from limited laboratory investigation reported in the literature. Similar studies can be conducted to fill the knowledge gap in cases where available analytical and experimental studies are limited or extend the information to cases where reliability information is unavailable. The study highlighted how reliability information can be extracted with the help of multiple deterministic analyses. As the secondary title of the book suggests, the authors believe that they proposed an alternative to the classical MCS technique.

9.2 Background Information

Solder is an easily meltable alloy commonly used to join two harder metals together in EP. The use of Tin-Lead (Sn-Pb) solder was common practice for decades. Information on its reliability and life cycle testing results are reported in the literature. However, lead (Pb) is categorized as a hazardous substance due to its toxic properties. At present, alternatives to lead-free solder alloys are under investigation. Tin is the most common element of the alloy. Adding copper and silver is expected to increase its resistance to thermal cyclic fatigue. Besides, copper can bring down the melting point and silver can increase the strength of the alloy. Due to their desirable characteristics, lead-free Tin-Silver-Copper solders are commonly used in the EP industry. However, information on their lifetime behavior and reliability is not yet comprehensively studied or reported in the literature. To address this knowledge gap, the underlying reliability of the Sn-3.9Ag-0.6Cu (SAC) lead-free solders used in EP under cyclic thermomechanical loading is presented in this chapter. To extract reliability information, the solder ball system shown in Figure 9.1, needs to be represented as realistically as practicable considering the physical and material behavior and then all major sources of uncertainty need to be incorporated in the formulation to estimate the underlying risk. The treatment of uncertainties in the deterministic formulation is emphasized here. Similar discussions can be found in Azizsoltani and Haldar (2018).

To meet these objectives, the basic system shown in Figure 9.1 is represented by FEs. This will be discussed in more detail later. The differences in coefficients of the thermal expansion between silicon, solder alloy, and copper substrate can result in thermomechanical stresses causing microcracks in solder balls inside an EP reducing its lifetime. Such microcracks can reduce the functionality of the solder balls to allow the flow of electrical current, serve as physical mechanical support for the microprocessor, and fail to manage or dissipate excessive heat.

Figure 9.1 Solder ball schematic.

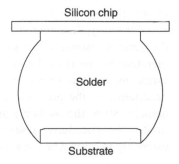

To capture the physics-based behavior of such a complex system consisting of substrate, solder ball, and silicon chip, it is necessary to capture the realistic material behavior of SAC by advanced material modeling techniques. To address the related issues, several constitutive modeling techniques with various degrees of sophistication are reported in the literature including hierarchical single-surface (HISS) (Desai et al. 1986) plasticity models with the distributed state concept (DSC) (Desai 2001, 2015), critical state continuous yielding models (Higgins et al. 2013), cap plasticity models (Motamedi and Foster 2015), and anisotropic damage plastic constitutive models (Azizsoltani et al. 2014; Khaloo et al. 2014). To capture the continuous yielding behavior, the HISS models are generally considered to be the most appropriate. Both critical state and cap plasticity models have some inherent weaknesses. They neglect the contribution of the deviatoric part of the plastic strains in yield surfaces and depend only on the volumetric plastic strains. They cannot model the non-associate material response and phase change from compression to dilation. Also, the material strength in such models is independent of the stress path and the yield surfaces are always circular in the principal stress space (Desai and Whitenack 2001). HISS can overcome these flaws. The HISS plasticity model with DSC is used in formulating the material behavior in developing the deterministic part of the work presented here.

A SAC solder ball will be subjected to thermomechanical loading. During normal operation, a SAC will experience an increase in the temperature with time. Thus, it will be necessary to apply thermomechanical loading in the time domain. Experimental studies show that the mechanical properties of solder alloys are highly dependent on the strain rate and temperature. Thus, solder alloys are expected to exhibit nonlinear viscoplastic deformations even at room temperature. The information available in the literature indicates that a deterministic representation of such a complicated system subject to thermomechanical loading will be very challenging due to the presence of several sources of uncertainties in the loading, physical dimensions, and material properties. It is essential that all major sources of uncertainty are appropriately incorporated in the deterministic formulation so that the underlying risk can be estimated accurately. If the information is judicially used, it will evaluate the overall performance of a solder ball. Obviously, the underlying risk should be kept as low as practicable; however, the accurate estimation of such risk can be very challenging. The absence of literature on the related topics justifies the statement. REDSET can be used to extract information on risk by considering all major sources of nonlinearity and uncertainty in the presence of thermomechanical loading applied in the time domain. Since the available analytical and experimental test results are very limited, the procedures presented here will help study the behavior of solder balls extending to cases where no similar study was conducted or reported.

9.3 Deterministic Modelling of a Solder Ball

As in any engineering study, an accurate deterministic formulation of the problem under consideration is a necessity. The study needs to incorporate information on geometric arrangements, material behavior, and loading conditions the system is expected to be exposed to during the normal operation. These three items are discussed separately in the following sections.

9.3.1 Solder Ball Represented by Finite Elements

Finite element representation of a complete ball grid array (BGA) surface-mount package subjected to thermomechanical cyclic loading is expected to be very computationally demanding. To identify the most critically stressed solder ball in the array, Zwick and Desai (1999) performed a linear elastic three-dimensional (3-D) thermomechanical analysis for a BGA package to locate the solder ball with the most severe stresses to reduce the physical size of the problem. Based on the results of that study, Whitenack (2004) concluded that a corner solder ball in BGA 225 surface-mount packaging would be the most critically stressed and is considered here for further study.

A two-dimensional (2-D) finite element representation of a corner solder ball in the array is shown in Figure 9.2. The geometric dimensions, as reported in Sane (2007) are used but the finite element mesh is developed using ABAQUS. For a 2-D representation of a 3-D solder ball, 360 four-node quadrilateral plane stress elements with 399 nodes, as shown in Figure 9.2 are considered. In this representation, the nodal thermal displacements were applied at the top between node points 379 and 399 and at the bottom between node points 3 and 19. The details of the thermomechanical loadings will be discussed in more detail later.

9.3.2 Material Modeling of SAC Alloy

It is necessary to incorporate appropriate material behavior into the elements at this stage. Constitutive modeling is the mathematical description of how materials behave under the applied loading. Under cyclic thermomechanical loading, the material goes through various phases before failure including elastic, plastic, creep deformation, softening, generation, and propagation of microcracks leading to fracture (Tucker et al. 2014). To develop working deterministic models incorporating information on these phases, information available in the literature is used. For ready reference, the information is summarized in the following sections (Azizsoltani and Haldar 2018).

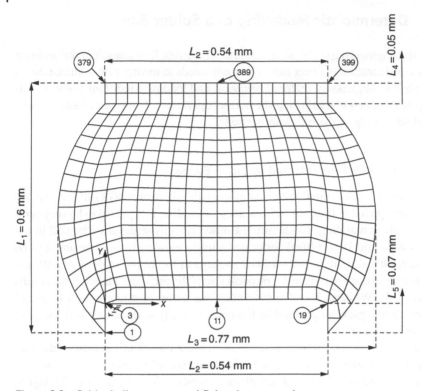

Figure 9.2 Solder ball geometry and finite element mesh.

9.3.2.1 HISS Plasticity Model

The HISS Constitutive model describes the realistic elastoplastic response behavior for several materials. In fact, the classical plasticity models such as von Misses, Tresca, Drucker-Prager, and Mohr-Coulomb can be derived from the HISS Constitutive model. The continuous HISS yield function, F, can be expressed as (Desai 2001):

$$F = \overline{J_{2D}} - \left(-\overline{\alpha} \overline{J_1}^n + \gamma \overline{J_1}^2 \right) \left(1 - \overline{\beta} S_r \right)^{-0.5} = 0 \tag{9.1a}$$

$$\overline{J_{2D}} = J_{2D}/p_a^2 \tag{9.1b}$$

$$\overline{J_1} = (J_1 + 3R)/p_a \tag{9.1c}$$

where J_{2D} is the second deviatoric stress invariant, J_1 is the first invariant of the stress tensor, p_a is the atmospheric pressure constant, $3R$ is the origin shift function which is the intercept of the HISS yield surface along J_1, $\overline{\alpha}$ is the hardening function, γ is the ultimate parameter, $\overline{\beta}$ is yield surface shape parameter, n is the phase change (compaction to dilation) parameter, and S_r is the stress ratio function. The origin shift function, $3R$, is defined as:

$$3R = \frac{\bar{c}}{\sqrt{\gamma}} \tag{9.2}$$

where $\bar{c}$ is the cohesive strength parameter.

Stress ratio function, S_r, is defined as:

$$S_r = \frac{\sqrt{27}}{2} \frac{J_{3D}}{J_{2D}{}^{3/2}} \tag{9.3}$$

where J_{3D} is the third deviatoric stress invariant.

The hardening function, $\bar{\alpha}$, is defined as:

$$\bar{\alpha} = \frac{a_1}{\xi^{\eta_1}} \tag{9.4}$$

where a_1 and η_1 are the hardening parameters, and ξ is the rate of plastic strains.

The rate of plastic strains, ξ, represents the accumulated plastic strain, and it is defined as:

$$\xi = \sqrt{d\varepsilon_{ij}^p \, d\varepsilon_{ij}^p} \tag{9.5}$$

where $d\varepsilon_{ij}^p$ is the increment of plastic strains, ε_{ij}^p.

The graphical representation of the HISS yield surfaces in J_1–J_{2D} space is shown in Figure 9.3 (Desai et al. 2011).

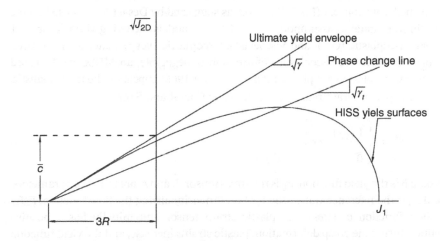

Figure 9.3 HISS yield surface in $J_1 - \sqrt{J_{2D}}$ space.

9.3.2.2 Disturbed State Concept

The DSC model assumes that the observed response of a solid is a mixture of a continuum part and a disturbed part (Desai 2016). For the continuum part, generally denoted as the relatively intact (RI) state, the classical elastoplastic material models can be used. The HISS model discussed in the previous section is used to represent the RI state in this study. The disturbed part is known as the fully adjusted (FA) state. The FA state is expected to contain microcracks and is considered to have just enough strength to carry the hydrostatic stress.

Using the disturbance function, D, Desai (2001) proposed the observed response of a deforming material in terms of RI and FA states as:

$$D = D_u \left(1 - e^{-A\xi_D{}^Z} \right) \tag{9.6}$$

where D_u is the ultimate disturbance parameter, A and Z are material parameters, and ξ_D is the rate of deviatoric plastic strain. The increment of the stress tensor $d\sigma_{ij}$ for the two material states can be expressed as (Desai 2001):

$$d\sigma_{ij}^o = (1-D)C_{ijkl}^i d\varepsilon_{kl}^i + DC_{ijkl}^f d\varepsilon_{kl}^f + dD\left(\sigma_{ij}^f - \sigma_{ij}^i\right) \tag{9.7}$$

where superscripts o, i, and f denote the observed, RI, and FA states, respectively, $d\varepsilon_{kl}$ is the incremental strain tensor, C_{ijkl} is the elastoplastic tangent operator, and dD is the incremental disturbance.

9.3.2.3 Creep Modeling

Creep is a time-dependent deformation of the material under the influence of stress. To consider the effect of both primary and secondary creeps, the multicomponent disturbed state concept (MDSC) is used, as suggested by Desai (2015). Based on the continuum material assumption, several creep models including viscoelastic (ve), elasto-viscoplastic (evp), and viscoelastic-viscoplastic (vevp) materials have been proposed. Since the viscoelastic deformation is negligible, an MDSC model based on Perzyna's elasto-viscoplastic model (Perzyna 1966) appears to be more realistic. The viscoplastic strain rate, $\dot{\varepsilon}_{ij}^{vp}$, is defined as (Desai and Sane 2013):

$$\dot{\varepsilon}_{ij}^{vp} = \begin{cases} \Gamma\left(\frac{F}{F_0}\right)^N \dfrac{\partial F}{\partial \sigma_{ij}}, & F > 0 \\ 0, & F \le 0 \end{cases} \tag{9.8}$$

where F is the yield function, σ_{ij} is the stress tensor, Γ and N are material parameters, and F_0 is the reference value used for nondimensionalizing the flow function, F/F_0. F is a function of stress and plastic strain tensor and initially has a positive value. During the creep deformation, plastic strains increase, and the yield function asymptotically approaches zero.

9.3.2.4 Rate-Dependent Elasto-Viscoplastic Model

Material responses are rate-dependent in most cases. The HISS model discussed earlier needs to be modified to consider the rate dependency. In order to make the HISS yield function rate dependent, the suggestion made by Baladi and Rohani (1984) is used. Dynamic plastic yield function, F_d, is defined as (Desai et al. 2011):

$$F_d(\boldsymbol{\sigma}, \bar{\alpha}, \dot{\varepsilon}) = \frac{F_S(\boldsymbol{\sigma}, \bar{\alpha}) - F_R(\dot{\varepsilon})}{F_0} \tag{9.9}$$

where subscripts d, R, and S represent dynamic, rate-dependent and static cases, respectively, $\boldsymbol{\sigma}$ is the stress tensor, $\dot{\varepsilon}$ is the rate of strain tensor, and other variables were defined earlier (Desai and Sane 2013).

9.3.3 Temperature-Dependent Modeling

The mechanical properties of SAC alloy are highly temperature dependent. Vianco et al. (2003) performed several stress-strain tests at different temperatures ranging from $-25\,°C$ to $160\,°C$ to investigate the sensitivity of mechanical properties of SAC alloy to temperature. Xiao and Armstrong (2005) conducted several creep tests at various temperatures ranging from $45\,°C$ to $150\,°C$. The results indicate that the modulus of elasticity, E, cohesive strength parameter, $\bar{c}$, hardening parameters, a_1 and η_1, and creep material parameters, Γ and N are temperature dependent. They need to be modeled accordingly to capture the physics-based behavior of the SAC alloy. The temperature dependency can be incorporated into the formulation using the relationship suggested by Sane (2007) as:

$$M = M_r \left(\frac{T}{T_r}\right)^C \tag{9.10}$$

where M can be any parameter, M_r is the parameter value at the reference temperature, T and T_r are the temperature of interest and the reference temperature in Kelvin, respectively, and C is the corresponding thermal exponent of each parameter. In this study, the reference temperature is considered to be $25\,°C$ for all temperature-dependent parameters identified earlier.

9.3.4 Constitutive Modeling Calibration

Thermal mismatches between the substrate, solder, and silicon chip material are considered to be the main sources of stress in solder balls. The coefficients of thermal expansion for different parts of a solder ball are shown in Table 9.1 (Tummala and Rymaszewski 1997).

Sane (2007) calibrated the material parameters for the HISS-DSC model using several types of experimental test data reported in the literature. Elastic, plastic,

Table 9.1 Coefficient of thermal expansion for solder ball materials.

Part	Material	Coefficient of thermal expansion (per million per Kelvin)
Silicon chip	Silicon	2.8
Solder	Sn-3.9Ag-0.6Cu	24
Substrate	Copper	16

Source: Adapted from Tummala and Rymaszewski (1997).

and hardening-related model parameters were calibrated using temperature-dependent stress-strain test results gathered from Vianco et al. (2003). Disturbance-related model parameters were calibrated using cyclic test data reported by Zeng et al. (2005). Creep-related model parameters were calibrated using creep test data in various temperatures reported by Xiao and Armstrong (2005). Details of all the parameters used in the constitutive modeling, their type and notation, their uses in the equations in the paper, and the sources used to generate information on them are summarized in Table 9.2.

9.3.5 Thermomechanical Loading Experienced by Solder Balls

The substrate and silicon sides of the solder ball undergo thermal displacements due to the difference in the coefficients of thermal expansion. Sane (2007) reported the nodal thermal displacements in the X and Y coordinates with respect to node number 3. The information is summarized in Table 9.3 and shown in Figure 9.2.

Both thermal displacement and thermal loading are applied in the cyclic form as shown in Figure 9.4 with the ramp time, T_1, of 18 minutes, the dwell time, T_2, of 30 minutes in high temperature, and idle time, T_3, of 10 minutes at low temperature. Thermal loading starts at 20 °C (Sane 2007). Then, the temperature is reduced to −55 °C and cycled between 125 °C and −55 °C. To incorporate the variation or uncertainty in the intensity of the thermal displacement and thermal loading, two magnification factors, denoted as MF_{TD} and MF_T, respectively, are introduced, as shown in Figure 9.4. MF_{TD}, MF_T, T_1, T_2, and T_3 are all considered to be random variables with different statistical characteristics, as will be discussed later. Figure 9.4 shows the time-dependent displacement and temperature variations for the first two cycles.

The FE representation of a solder ball, material behavior, and thermomechanical loading required to develop a deterministic model are available at this stage. It is now necessary to incorporate uncertainties in all the random variables in the formulation to extract reliability information.

Table 9.2 Model parameter calibration procedure.

Test	Source	Type	Parameter	Parameter description	Related Eqs.
Temperature-dependent stress-strain tests	Vianco et al. (2003)	Elastic	E	Elastic Modulus	Hooke's law
			ν	Poisson's ratio	Hooke's law
		Plastic	γ	Ultimate parameter	9.1a
			$\bar{\beta}$	Yield surface shape parameter	9.1a
			n	Phase change parameter	9.1a
			$\bar{c}$	Cohesive strength parameter	9.2
		Hardening	a_1	Hardening parameter	9.4
			η_1	Hardening parameter	9.4
Cyclic tests	Zeng et al. (2005)	Disturbance	D_u	Ultimate disturbance parameter	9.6
			A	Disturbance parameter	9.6
			Z	Disturbance parameter	9.6
Creep tests	Xiao and Armstrong (2005)	Creep	Γ	Fluidity parameter	9.8
			N	Flow function exponent	9.8

9.4 Uncertainty Quantification

With the availability of the deterministic model for the solder balls and the thermomechanical loading they are subjected to, it is required to quantify uncertainty in all the random variables in the formulation. Uncertainties in the solder balls and the thermomechanical loading are discussed separately next.

Table 9.3 Thermal displacements with respect to node number 3.

Node number	X Displacement (mm)	Y Displacement (mm)	Node number	X Displacement (mm)	Y Displacement (mm)
4	8.83920E−5	4.11480E−5	399	−7.16280E−4	2.32410E−3
5	1.76784E−4	8.20420E−5	398	−6.52780E−4	2.36220E−3
6	2.64160E−4	1.23190E−4	397	−5.89280E−4	2.39776E−3
7	3.53060E−4	1.64084E−4	396	−5.89280E−4	2.43332E−3
8	4.41960E−4	2.05232E−4	395	−4.59740E−4	2.47142E−3
9	5.30860E−4	2.46126E−4	394	−3.96240E−4	2.50698E−3
10	6.19760E−4	2.87020E−4	393	−3.32740E−4	2.54000E−3
11	7.06120E−4	3.27660E−4	392	−2.69240E−4	2.59080E−3
12	7.95020E−4	3.68300E−4	391	−2.05486E−4	2.61620E−3
13	8.83920E−4	4.11480E−4	390	−1.41732E−4	2.64160E−3
14	9.72820E−4	4.52120E−4	389	−7.79780E−5	2.69240E−3
15	1.06172E−3	4.92760E−4	388	−1.41478E−5	2.71780E−3
16	1.14808E−3	5.33400E−4	387	4.95300E−5	2.76860E−3
17	1.23698E−3	5.74040E−4	386	1.13538E−4	2.79400E−3
18	1.32588E−3	6.14680E−4	385	1.77292E−4	2.84480E−3
19	1.41478E−3	6.55320E−4	384	2.41046E−4	2.87020E−3
			383	3.04800E−4	2.92100E−3
			382	3.68300E−4	2.94640E−3
			381	4.31800E−4	2.97180E−3
			380	4.95300E−4	3.02260E−3
			379	5.58800E−4	3.04800E−3

9.4.1 Uncertainty in all the Parameters in a Solder Ball

As discussed earlier, statistical information on all the RVs present in the deterministic model of a solder ball is not available similar to what is available for a typical structure. The required information was generated in a very methodical way. Sane (2007) considered all the variables deterministic in his study. The authors considered them as the mean values. Then, the authors considered the coefficient of variation (COV) and the statistical distribution using the information on similar variables used in civil engineering applications and their experience dealing with them as best as possible. The information is summarized in Table 9.4.

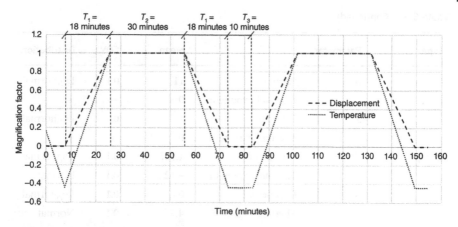

Figure 9.4 Variation of MF_{TD}, MF_T.

Table 9.4 Statistical information of RVs.

Type	Parameter		Mean	COV	Distribution
Elastic	E (Mpa)	Parameter	2343	0.06	Lognormal
		Thermal exponent	−1.12	0.1	Normal
	ν	Parameter	0.3474	0.1	Lognormal
		Thermal exponent	0	—	—
Plastic	γ	Parameter	8.35E−4	0.1	Lognormal
		Thermal exponent	0	—	—
	$\bar{\beta}$	Parameter	0	—	—
		Thermal exponent	0	—	—
	n	Parameter	2.1	0.1	Lognormal
		Thermal exponent	0	—	—
	$\bar{c}$	Parameter	5.28	0.1	Lognormal
		Thermal exponent	−3.2	0.1	Normal
Hardening	a_1	Parameter	6E−8	0.1	Lognormal
		Thermal exponent	−3.1	0.1	Normal
	η_1	Parameter	2.5	0.1	Lognormal
		Thermal exponent	0.2	0.1	Normal
Disturbance	D_u	Parameter	1	0.1	Lognormal
		Thermal exponent	0	—	—

(Continued)

Table 9.4 (Continued)

Type	Parameter		Mean	COV	Distribution
	A	Parameter	11	0.1	Lognormal
		Thermal exponent	0	—	—
	Z	Parameter	1	0.1	Lognormal
		Thermal exponent	0	—	—
Creep	Γ	Parameter	1.16E−9	0.1	Lognormal
		Thermal exponent	−12.3	0.1	Normal
	N	Parameter	0.4	0.1	Lognormal
		Thermal exponent	4.7	0.1	Normal
Dimension		L_1 (mm)	0.6	0.05	Lognormal
		L_2 (mm)	0.54	0.05	Lognormal
		L_3 (mm)	0.77	0.05	Lognormal
		L_4 (mm)	0.05	0.05	Lognormal
		L_5 (mm)	0.07	0.05	Lognormal
Thickness		TH (mm)	0.54	0.05	Lognormal
Load RVs		T_1 (min)	18	0.1	Type 1
		T_2 (min)	30	0.1	Type 1
		T_3 (min)	10	0.1	Type 1
		MF_T	1	0.1	Type 1
		MF_{TD}	1	0.1	Type 1

9.4.2 Uncertainty Associated with Thermomechanical Loading

As discussed in Section 9.3.5, the thermomechanical loading is described in terms of MF_{TD}, MF_T, T_1, T_2, and T_3. They are considered to be RVs with Type 1 Gumbel distribution with a COV of 0.1.

9.5 The Limit State Function for the Reliability Estimation

The required limit state function (LSF) for solder balls under thermomechanical loading is expressed in terms of the volume fraction of the material (V_f) that reaches a critical disturbance (D_f) set by the designer. The concept can be mathematically expressed as:

$$g(\mathbf{X}) = V_{f\,\text{allow}} - V_{fD \geq D_f} = V_{f\,\text{allow}} - \hat{g}(\mathbf{X}) \tag{9.11}$$

where $V_{f\text{allow}}$ is the allowable disturbed volume fraction of the material and $V_{fD \geq D_f}$ is the disturbed volume fraction obtained by FE analyses. Sane (2007) proposed thresholds for D_f and $V_{f\text{allow}}$ as 0.9 and 0.025, respectively. Simply stated, when 2.5% of the volume of the material reaches a 90% disturbance level, a solder ball can be considered to be in the failure state.

9.6 Reliability Assessment of Lead-Free Solders in Electronic Packaging

With the availability of all the required information and uncertainty quantification, it is now possible to extract reliability information on the most critically stressed solder ball using REDSET. The required IRS and the corresponding LSF are generated by REDSET using Sampling Scheme MS3-KM. At the initiation of the risk evaluation process, the total number of RVs for the solder ball shown in Figure 9.2 is found to be $k = 29$. After the sensitivity analysis, k_R is selected to be 8. To implement Sampling Scheme MS3-KM, m is found to be 4. The information available in the literature indicates that the solder ball shown in Figure 9.2 failed at 3248 cycles of thermomechanical loading during laboratory testing. This prompted the authors to subject the solder ball with about 80% of 3248 or about 2600 cycles of thermomechanical loading in the presence of uncertainties summarized in Table 9.4 to study the capability of REDSET.

The probability of failure p_f of the solder ball is then estimated using REDSET with Sampling Scheme MS3-KM. The results are summarized in Table 9.5. In Eq. 3.10, an arbitrary factor h_i was used to define the experimental region. To study the effect of the experimental region and denoting h_i^i and h_i^f are the values of the

Table 9.5 Verification using 2600 cycles of thermomechanical loading.

LSF		Proposed MKM	MCS
$h^i = 1$ $h^f = 1$	β	1.7080	1.7340
	p_f (E−2)	4.3821	4.1459
	NDFEA	109	10000
$h^i = 1$ $h^f = 0.5$	β	1.7275	1.7340
	p_f (E−2)	4.2038	4.1459
	NDFEA	109	10000

h_i parameter for intermediate and the final iterations, respectively, the p_f values are estimated for two different values of h_i^f and are summarized in the table. When h_i^f values are 0.5 and 1.0, the corresponding p_f values are about 0.042 and 0.044, respectively, indicating a slightly smaller probability of failure for the smaller size of the experimental region.

9.7 Numerical Verification Using Monte Carlo Simulation

The capability of REDSET in estimating p_f needs to be verified at this stage. The basic MCS technique is for this purpose. To address the issue related to the total number of simulations required for the verification, the suggestion made by Haldar and Mahadevan (2000a) is used. When the LSF represented by Eq. 9.11 is less than zero, it will indicate failure. Suppose N_f is the number of simulation cycles when Eq. 9.11 is less than zero and N is the total number of simulation cycles. Then an estimate of p_f can be expressed as:

$$p_f = \frac{N_f}{N} \tag{9.12}$$

Denoting p_f^T is the actual probability of failure, the percentage error, $\varepsilon\%$, in estimating it can be expressed as:

$$\varepsilon\% = \frac{\frac{N_f}{N} - p_f^T}{p_f^T} \times 100\% \tag{9.13}$$

By combining Eq. 9.13 with the 95% confidence interval of the estimated probability of failure, Haldar and Mahadevan (2000a) suggested that

$$\varepsilon\% = \sqrt{\frac{\left(1 - p_f^T\right)}{N \times p_f^T}} \times 200\% \tag{9.14}$$

Equation 9.14 indicates that there will be about a 20% error if p_f^T is 0.01 and 10 000 trials were used in the simulation. It can also be interpreted as that there will be a 95% probability that p_f will be in the range of 0.01 ± 0.002 with 10 000 simulations. In order to estimate p_f of the order of 0.04 with about 10% error, the total number of simulations will be around 10 000. Thus, 10 000 simulations were used to verify the results obtained by REDSET.

It requires about 9028 hours of continuous running of a computer. As mentioned earlier, the task was executed using the advanced computing facilities available at the University of Arizona and the parallel processing technique. The results obtained by MCS are also summarized in Table 9.5. The results clearly indicate the information on p_f obtained by REDSET Sampling Scheme MS3-MK and MCS are similar indicating the proposed method is verified. However, the total number of nonlinear dynamic finite element analyses (TNDFEA) is only 109 instead of 10 000 required by MCS. The results also indicate that when the simulation region is reduced, the estimated probability of failure using REDSET is closer to the simulation. The proposed method is very accurate and efficient.

9.8 Verification Using Laboratory Test Results

The successful analytical verification as presented in the previous section prompted the authors to check the predictability of estimating p_f of the SAC solder ball shown in Figure 9.2. The p_f values under several different thermomechanical loading cycles are estimated using REDSET with Sampling Scheme MS3-KM. The reliability index, β, the corresponding p_f value, and the required TNDFEA for each loading cycle are summarized in Table 9.6. The results are also plotted in Figure 9.5. As expected, the p_f value increased with

Table 9.6 PF for different cycles of thermomechanical loading using MKM.

Cycle number	β	p_f	TNDFEA
3500	−3.5441	0.9998	125
3400	−2.3841	0.9914	125
3300	−1.9724	0.9757	109
3200	−1.1850	0.8820	101
3100	−0.5478	0.7081	101
3000	0.0854	0.4660	101
2900	0.4671	0.3202	101
2800	0.8558	0.1960	109
2700	1.2530	0.1051	109
2600	1.7275	0.0420	109
2500	1.9307	2.6763E−2	109
2400	2.1809	1.4594E−2	109
2300	2.5788	4.9574E−3	125
2200	2.6133	4.4834E−3	125
2100	2.9962	1.3668E−3	125

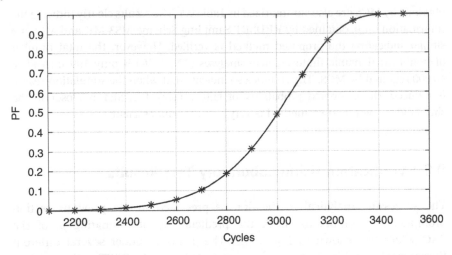

Figure 9.5 Probability of failure vs. number of thermomechanical loading cycles profile.

the increase in the number of loading cycles. It is only 1.3668E−3 with 2100 cycles of loading, but very close to 1.00 at 3500 cycles. During the laboratory investigation, the solder failed at 3248 cycles [private communication, as reported by Sane (2007)]. It can be observed from Figure 9.5 that REDSET with Sampling Scheme MS3-KM predicted the analytical p_f value of about 0.9238 at 3248 cycles. The nature of the curve in Figure 9.5 also indicates the appropriate prediction behavior of REDSET with Sampling Scheme MS3-KM and can be used in predicting the life expectancy of solder balls. The results of the p_f value versus the number of thermomechanical loading cycles can be important for sensitive applications in aerospace and medical areas.

In summary, considering both the analytical and laboratory experimental results, REDSET is now conclusively verified. It is capable of estimating the reliability of solder balls.

9.9 Concluding Remarks

The reliability evaluation of a lead-free solder ball used in EP subjected to thermo-mechanical loading is presented. A solder ball is represented by finite elements and major sources of nonlinearities are incorporated. Uncertainties in all design variables are quantified. The thermomechanical loading is represented by five design parameters and uncertainties associated with them are incorporated. The accuracy, efficiency, and application potential of the procedure are established with the help of MCS and results from laboratory investigations reported in the literature. The study conclusively verified REDSET with Sampling Scheme MS3-KM. Similar studies can be conducted to fill the knowledge gap in cases where available analytical and experimental studies are limited or extend the information to cases where reliability information is unavailable. The study showcased how reliability information can be extracted with the help of multiple deterministic analyses. The author believes that they proposed an alternative to the classical MCS technique.

Concluding Remarks for the Book - REDSET

Reliability analysis with implicit limit state functions has always been challenging and difficult, particularly when nonlinear dynamic systems are excited in the time domain. A novel reliability analysis procedure denoted as REDSET is presented in this book to address this vacuum. To address uncertainty in the dynamic excitation applied in the time domain, multiple time histories need to be selected for design purposes. In the United States at present, at least 7 need to be selected to satisfy predefined criteria. This requirement is a step in the right direction but fails to consider other sources of uncertainty in engineering systems. A relatively new design concept currently under development in the United States is known as Performance Based Seismic Design (PBSD). It is a very sophisticated risk-based design guideline. However, no acceptable risk-based design procedure was suggested in the development process of PBSD, and no method is currently available to implement it. REDSET is proposed to fill this vacuum. Instead of conducting 7 nonlinear finite dynamic finite element (FE) analyses, comprehensive information on the underlying risk can be extracted with the help of REDSET by conducting a few dozen or a few hundred deterministic nonlinear dynamic FE analyses. Considering the enormous computational power that is currently available, any computer program in different computational platforms can be used to generate the information on the underlying risk with few additional FE analyses. Structural systems can be represented by the displacement-based or the assumed stress-based FE method. REDSET is extensively verified for different engineering systems including onshore and offshore structures, and solder balls used in computer chips. Dynamic excitations applied in the time domain include earthquakes, wind-induced waves, and the thermomechanical loading caused by heating and cooling. The advanced capabilities of REDSET are demonstrated with the help of many examples in the book. Some of its advanced features (currently not available to engineers) are expected to satisfy specified design requirements. REDSET is shown to be efficient, accurate, and robust. The evidence provided in the book may convince the readers to use REDSET for the reliability estimation of CNDES excited in the time domain in the future.

As the secondary title of the book suggests, REDSET can be considered as an alternative to the classical random vibration concept and the basic simulation procedure for risk estimation purposes. It changes the current engineering design paradigm. Instead of conducting one deterministic analysis, a design can be made more dynamic load tolerant, resilient, and sustainable with the help of a few additional deterministic analyses using REDSET.

References

AISC 341-10 (2010). *Seismic Provisions for Structural Steel Buildings*. Chicago, IL: American Institute of Steel Construction (AISC).

American Institute of Steel Construction (AISC) (2011, 2017). *Steel Construction Manual*. Chicago, IL: American Institute of Steel Construction.

American Petroleum Institute (API) (2007). *Recommended Practice for Planning, Designing and Constructing Fixed Offshore Platforms – Working Stress Design. RP 2A-WSD*, 21e. Washington, DC: American Petroleum Institute.

ASCE/SEI 7-10 (2010). *Minimum Design Loads for Buildings and Other Structures*. Reston, VA: American Society of Civil Engineers (ASCE).

ASCE/SEI 41 (2014, 2017). *Seismic Evaluation and Retrofit of Existing Buildings*. American Society of Civil Engineers (ASCE), Reston, VA, USA.

ATC 3-06 (1978), *Tentative Provisions for the Development of Seismic Regulations for Buildings*. Applied Technology Council (ATC), Redwood City, California, USA.

Ayyub, B.M. and Haldar, A. (1984). Practical structural reliability techniques. *Journal of Structural Engineering* 110 (8): 1707–1724. https://doi.org/10.1061/(ASCE)0733-9445 (1984)110:8(1707).

Azizsoltani, H. (2017). Risk estimation of nonlinear time domain dynamic analyses of large systems. PhD thesis. University of Arizona.

Azizsoltani, H. and Haldar, A. (2017). A surrogate concept of multiple deterministic analyses of non-linear structures excited by dynamic loadings. *Proceedings of the 12th International Conference on Structural Safety and Reliability*, Vienna, Austria.

Azizsoltani, H. and Haldar, A. (2018). Reliability analysis of lead-free solders in electronic packaging using a novel surrogate model and kriging concept. *Journal of Electronic Packaging* 140 (4): 041003. https://doi.org/10.1115/1.4040924.

Azizsoltani, H., Kazemi, M.T., and Javanmardi, M.R. (2014). An Anisotropic damage model for metals based on irreversible thermodynamics framework. *Iranian Journal of Science and Technology. Transactions of Civil Engineering* 38 (C1+): 157. https://doi.org/10.22099/ijstc.2014.1861.

Reliability Evaluation of Dynamic Systems Excited in Time Domain: Alternative to Random Vibration and Simulation, First Edition. Achintya Haldar, Hamoon Azizsoltani, J. Ramon Gaxiola-Camacho, Sayyed Mohsen Vazirizade, and Jungwon Huh.
© 2023 John Wiley & Sons, Inc. Published 2023 by John Wiley & Sons, Inc.

Azizsoltani, H., Gaxiola-Camacho, J.R., and Haldar, A. (2018). Site-specific seismic design of damage tolerant structural systems using a novel concept. *Bulletin of Earthquake Engineering* 16 (9): 3819–3843. https://doi.org/10.1007/s10518-018-0329-5.

Baladi, G.Y. and Rohani (1984). Development of an elastic-viscoplastic constitutive relationship for earth materials. *Mechanics of Engineering Materials* 23–43.

Bathe, K.J. (1982). *Finite Element Procedures in Engineering Analysis*. Englewood Cliffs, NJ: Prentice-Hall.

Beyer, K. and Bommer, J.J. (2007). Selection and scaling of real accelerograms for bi-directional loading: a review of current practice and code provisions. *Journal of Earthquake Engineering* 11 (S1): 13–45. https://doi.org/10.1080/13632460701280013.

Bhattacharjya, S. and Chakraborty, S. (2011). Robust optimization of structures subjected to stochastic earthquake with limited information on system parameter uncertainty. *Engineering Optimization* 43 (12): 1311–1330. https://doi.org/10.1080/0305215X.2011.554545.

Bitner-Gregersen, E. M. (2010). Uncertainties of Joint Long-Term Probabilistic Modelling of Wind Sea and Swell, 29th International Conference on Ocean, Offshore and Arctic Engineering, https://asmedigitalcollection.asme.org/OMAE/proceedings-abstract/OMAE2010/49101/493/359872.

Bitner-Gregersen, E. M. (2015). Joint met-ocean description for design and operations of marine structures. *Appl Ocean Res* 51: 279–292. https://www.sciencedirect.com/science/article/abs/pii/S0141118715000085.

Bitner-Gregersen, E.M. and Haver, S. (1989). Joint long term description of environmental parameters for structural response calculation. *Proceedings of the 2nd International Workshop on Wave Hindcasting and Forecasting*, Vancouver, B.C., 25–28 April(pp. 25–28). Downsview, Ont.: Environment Canada, Atmospheric Environment Service.

Bitner-Gregersen, E.M., Cramer, E.H., and Løseth, R. (1995). Uncertainties of load characteristics and fatigue damage of ship structures. *Marine Structures* 8 (2): 97–117. https://doi.org/10.1016/0951-8339(94)00013-I.

Bommer, J.J. and Acevedo, A.B. (2004). The use of real earthquake accelerograms as input to dynamic analysis. *Journal of Earthquake Engineering* 8 (spec01): 43–91. https://doi.org/10.1080/13632460409350521.

Box, M.J. and Draper, N.R. (1971). Factorial designs, the |X′X| criterion, and some related matters. *Technometrics* 13 (4): 731–742.

Box, G.E.P. and Wilson, K.B. (1951). On the experimental attainment of optimum conditions. *Journal of the Royal Statistical Society. Series B (Methodological)* 13 (1): 1–45. https://www.jstor.org/stable/2983966.

Box, G.E., Hunter, W.G., and Hunter, J.S. (1978). *Statistics for Experimenters: An Introduction to Design, Data Analysis, and Model Building*. New York: Wiley.

Bucher, C.G. and Bourgund, U. (1990). A fast and efficient response surface approach for structural reliability problems. *Structural Safety* 7 (1): 57–66. https://doi.org/10.1016/0167-4730(90)90012-E.

Bucher, C.G., Chen, Y.M., and Schuëller, G.I. (1989). Time variant reliability analysis utilizing response surface approach. *Reliability and Optimization of Structural Systems'*, (Vol. 88, pp. 1–14). Berlin, Heidelberg: Springer. https://doi.org/10.1007/978-3-642-83828-6_1.

Burks, L.S., Zimmerman, R.B., and Baker, J.W. (2015). Evaluation of hybrid broadband ground motion simulations for response history analysis and design. *Earthquake Spectra* 31 (3): 1691–1710. https://doi.org/10.1193/091113EQS248M.

Cacciola, P. and Deodatis, G. (2011). A method for generating fully non-stationary and spectrum-compatible ground motion vector processes. *Soil Dynamics and Earthquake Engineering* 31 (3): 351–360. https://doi.org/10.1016/j.soildyn.2010.09.003.

Cacciola, P. and Zentner, I. (2012). Generation of response-spectrum-compatible artificial earthquake accelerograms with random joint time–frequency distributions. *Probabilistic Engineering Mechanics* 28: 52–58. https://doi.org/10.1016/j.probengmech.2011.08.004.

Canadian Standard Association (CSA) (1974). *Standards for the Design of Cold-Formed Steel Members in Buildings*. CSA S-137.

Cassidy, M.J. (1999). Non-linear analysis of jack-up structures subjected to random waves. PhD thesis. University of Oxford.

Cassidy, M.J., Taylor, R.E., and Houlsby, G.T. (2001). Analysis of jack-up units using a constrained NewWave methodology. *Applied Ocean Research* 23 (4): 221–234. https://doi.org/10.1016/S0141-1187(01)00005-0.

Cassidy, M.J., Taylor, P.H., Taylor, R.E., and Houlsby, G.T. (2002). Evaluation of long-term extreme response statistics of jack-up platforms. *Ocean Engineering* 29 (13): 1603–1631. https://doi.org/10.1016/S0029-8018(01)00110-X.

Castillo, E. (1988). *Extreme Value Theory in Engineering*. San Diego, California: Academic Press.

Chakraborty, S. and Sen, A. (2014). Adaptive response surface based efficient finite element model updating. *Finite Elements in Analysis and Design* 80: 33–40.

Chen, W.F. and Kishi, N. (1989). Semirigid steel beam-to-column connections: data base and modeling. *Journal of Structural Engineering* 115 (1): 105–119. https://doi.org/10.1061/(ASCE)0733-9445(1989)115:1(105).

Chen and Lind (1983). Fast Probability integration by three-parameter normal tail approximation. *Structural Safety* 1: 269–276.

Chen, W.F. and Lui, E.M. (1987). Effects of joint flexibility on the behavior of steel frames. *Computers & Structures* 26 (5): 719–732. https://doi.org/10.1016/0045-7949(87)90021-6.

Clough, R.W. and Penzien, J. (1993). *Dynamics of Structures*, 2e. New York: McGraw-Hill.

Colson, A. (1991). Theoretical modeling of semirigid connections behavior. *Journal of Constructional Steel Research* 19 (3): 213–224. https://doi.org/10.1016/0143-974X(91)90045-3.

Comité European du Be'ton (CEB) (1976). Joint Committee on Structural Safety CEB-CECM-IABSE-RILEM First Order Reliability Concepts for Design Codes, *CEM Bulletin No. 112*.

Cressie, N. (2015). *Statistics for Spatial Data*. Wiley.

Dawson, T.H. (1983). *Offshore Structural Engineering*. Englewood Cliffs, NJ: Prentice-Hall.

Der Kiureghian, A. and Ke, J.B. (1985). Finite-element based reliability analysis of frame structures. *Proceedings of the 4th International Conference on Structural Safety and Reliability* (Vol. 1, pp. 395–404). New York: International Association for Structural Safety and Reliability.

Desai, C.S. (2001). *Mechanics of Materials and Interfaces: The Disturbed State Concept*. CRC Press.

Desai, C.S. (2015). Constitutive modeling of materials and contacts using the disturbed state concept: part 2–Validations at specimen and boundary value problem levels. *Computers & Structures* 146: 234–251. https://doi.org/10.1016/j. compstruc.2014.07.026.

Desai, C.S. (2016). Disturbed state concept as unified constitutive modeling approach. *Journal of Rock Mechanics and Geotechnical Engineering* 8 (3): 277–293. https://doi.org/10.1016/j.jrmge.2016.01.003.

Desai, C.S. and Sane, S.M. (2013). Rate dependent elastoviscoplastic model. In: *Constitutive Modeling of Geomaterials*, Springer Series in Geomechanics and Geoengineering (ed. Q. Yang, J.M. Zhang, H. Zheng and Y. Yao), 97–105. Berlin, Heidelberg: Springer https://doi.org/10.1007/978-3-642-32814-5_8.

Desai, C.S. and Whitenack, R. (2001). Review of models and the disturbed state concept for thermomechanical analysis in electronic packaging. *Journal of Electronic Packaging* 123 (1): 19–33. https://doi.org/10.1115/1.1324675.

Desai, C.S., Somasundaram, S., and Frantziskonis, G. (1986). A hierarchical approach for constitutive modelling of geologic materials. *International Journal for Numerical and Analytical Methods in Geomechanics* 10 (3): 225–257. https://doi.org/10.1002/ nag.1610100302.

Desai, C.S., Sane, S., and Jenson, J. (2011). Constitutive modeling including creep-and rate-dependent behavior and testing of glacial tills for prediction of motion of glaciers. *International Journal of Geomechanics* 11 (6): 465–476. https://doi.org/ 10.1061/(ASCE)GM.1943-5622.0000091.

Dyanati, M. and Huang, Q. (2014). Seismic reliability of a fixed offshore platform against collapse. *International Conference on Offshore Mechanics and Arctic Engineering* (Vol. 45431, p. V04BT02A020). American Society of Mechanical Engineers. https://doi.org/10.1115/OMAE2014-24137.

Eik, K.J. and Nygaard, E. (2003). *Statfjord Late Life Metocean Design Basis*. Stavanger: Statoil.

Ellingwood, B., Galambos, T.V., MacGregor, J.G., and Cornell, C.A. (1980). *Development of a Probability Based Load Criterion for American National Standard A58: Building Code Requirements for Minimum Design Loads in Buildings and Other Structures*, vol. 577. Department of Commerce, National Bureau of Standards.

Elsati, M.K. and Richard, R.M. (1996). Derived moment rotation curves for partially restrained connections. *Structural Engineering Review* 2 (8): 151–158. https://doi.org/10.1016/0952-5807(95)00055-0.

Engelhardt, M.D. and Sabol, T.A. (1998). Reinforcing of steel moment connections with cover plates: benefits and limitations. *Engineering Structures* 20 (4–6): 510–520. https://doi.org/10.1016/S0141-0296(97)00038-2.

Farag, R., Haldar, A., and El-Meligy, M. (2016). Reliability analysis of piles in multilayer soil in mooring dolphin structures. *Journal of Offshore Mechanics and Arctic Engineering* 138 (5): https://doi.org/10.1115/1.4033578.

FEMA P-58 (2012). *Seismic Performance Assessment of Buildings*. Washington, DC: Federal Emergency Management Agency (FEMA).

FEMA P-751 (2012). *NEHRP Recommended Seismic Provisions: Design Examples*. Washington, DC: Federal Emergency Management Agency (FEMA).

FEMA-273 (1997). *NEHRP Guidelines for the Seismic Rehabilitation of Buildings*. Washington, DC: Federal Emergency Management Agency (FEMA).

FEMA-350 (2000). *Recommended Seismic Design Criteria for New Steel Moment-Frame Buildings*. Washington, DC: Federal Emergency Management Agency (FEMA).

FEMA-351 (2000). *Recommended Seismic Evaluation and Upgrade Criteria for Existing Welded Steel Moment-Frame Buildings*. Washington, DC: Federal Emergency Management Agency (FEMA).

FEMA-352 (2000). *Recommended Post-earthquake Evaluation and Repair Criteria for Welded Steel Moment-Frame Buildings*. Washington, DC: Federal Emergency Management Agency (FEMA).

FEMA-353 (2000). *Recommended Specifications and Quality Assurance Guidelines for Steel Moment-Frame Construction for Seismic Applications*. Washington, DC: Federal Emergency Management Agency (FEMA).

FEMA-354 (2000). *A Policy Guide to Steel Moment Frame Construction*. Washington, D.C..

FEMA-355C (2000). *State of the Art Report on Systems Performance of Steel Moment Frames Subject to Earthquake Ground Shaking*. Washington, DC: Federal Emergency Management Agency (FEMA).

FEMA-355F (2000). *State of the Art Report on Performance Prediction and Evaluation of Steel Moment-Frame Buildings*. Washington, DC: Federal Emergency Management Agency (FEMA).

FEMA-P695 (2009). *Quantification of Building Seismic Performance Factors*. Washington, DC: Federal Emergency Management Agency (FEMA).

Forristall, G.Z. and Cooper, C.K. (1997). Design current profiles using empirical orthogonal function (EOF) and inverse FORM methods. *Offshore Technology Conference*, OnePetro. https://doi.org/10.4043/8267-MS.

Freudenthal, A.M. (1956). Safety and the probability of structural failure. *Transactions of the American Society of Civil Engineers* 121 (1): 1337–1375. https://doi.org/10.1061/TACEAT.0007306.

Frye, M.J. and Morris, G.A. (1975). Analysis of flexibly connected steel frames. *Canadian Journal of Civil Engineering* 2 (3): 280–291. https://doi.org/10.1139/l75-026.

Gavin, H.P. and Yau, S.C. (2008). High-order limit state functions in the response surface method for structural reliability analysis. *Structural Safety* 30 (2): 162–179. https://doi.org/10.1016/j.strusafe.2006.10.003.

Gaxiola-Camacho, J.R. (2017). Performance-based seismic design of nonlinear steel structures using a novel reliability technique. PhD thesis. The University of Arizona.

Ghobarah, A. (2001). Performance-based design in earthquake engineering: state of development. *Engineering Structures* 23 (8): 878–884. https://doi.org/10.1016/S0141-0296(01)00036-0.

Golafshani, A.A., Bagheri, V., Ebrahimian, H., and Holmas, T. (2011a). Incremental wave analysis and its application to performance-based assessment of jacket platforms. *Journal of Constructional Steel Research* 67 (10): 1649–1657. https://doi.org/10.1016/j.jcsr.2011.04.008.

Graves, R.W. and Pitarka, A. (2010). Broadband ground-motion simulation using a hybrid approach. *Bulletin of the Seismological Society of America* 100 (5A): 2095–2123. https://doi.org/10.1785/0120100057.

Gumbel, E.J. (1958). *Statistics of Extremes*. New York: Columbia University Press.

Haldar, A. (1981). Probabilistic evaluation of construction deficiencies. *Journal of the Construction Engineering Division, ASCE* 107 (CO1): 107–119.

Haldar, A. and Ayyub, B.M. (1984). Practical Variance Reduction Techniques in Simulation, *Advances in Probabilistic Structural Mechanics - 1984*, American Society of Mechanical Engineers, *PVP - 93*, 63–74.

Haldar, A. and Mahadevan, S. (2000a). *Probability, Reliability, and Statistical Methods in Engineering Design*. New York: Wiley.

Haldar, A. and Mahadevan, S. (2000b). *Reliability Assessment using Stochastic Finite Element Analysis*. New York: Wiley.

Hamburger, R.O. and Hooper, J.D. (2011). Performance-based seismic design. *Modern Steel Construction* 51 (4): 36–39.

Hamburger, R.O., Krawinkler, H., Malley, J.O., and Adan Scott, M. (2009), Seismic design of steel special moment frames, *NEHRP seismic design technical brief*, (2).

Hartzell, S., Harmsen, S., Frankel, A., and Larsen, S. (1999). Calculation of broadband time histories of ground motion: comparison of methods and validation using strong-ground motion from the 1994 Northridge earthquake. *Bulletin of the Seismological Society of America* 89 (6): 1484–1504. https://doi.org/10.1785/BSSA0890061484.

Hasofer, A.M. and Lind, N.C. (1974). Exact and invariant second-moment code format. *Journal of the Engineering Mechanics Division* 100 (1): 111–121. https://doi.org/10.1061/JMCEA3.0001848.

Hasselmann, K., Barnett, T.P., Bouws, E. et al. (1973). Measurements of wind-wave growth and swell decay during the Joint North Sea Wave Project (JONSWAP). *Ergaenzungsheft zur Deutschen Hydrographischen Zeitschrift, Reihe A* http://hdl.handle.net/21.11116/0000-0007-DD3C-E.

Haver, S. (1985). Wave climate off northern Norway. *Appl. Ocean Res* 7(2) 85–92, https://www.sciencedirect.com/science/article/abs/pii/0141118785900380.

Haver, S. and Winterstein, S.R. (2008). Environmental contour lines: a method for estimating long term extremes by a short term analysis. *SNAME Maritime Convention*, OnePetro. https://doi.org/10.5957/SMC-2008-067.

Hengl, T. (2009). *A Practical Guide to Geostatistical Mapping*, vol. 52. Tomislav Hengl.

Henry, Z., Jusoh, I., and Ayob, A. (2017). Structural Integrity Analysis of Fixed Offshore Jacket Structures. *Jurnal Mekanikal* 40: 23–36.

Higgins, W., Chakraborty, T., and Basu, D. (2013). A high strain-rate constitutive model for sand and its application in finite-element analysis of tunnels subjected to blast. *International Journal for Numerical and Analytical Methods in Geomechanics* 37 (15): 2590–2610. https://doi.org/10.1002/nag.2153.

Hinton, E. and Owen, D.R.J. (1986). *Finite Elements in Plasticity: Theory and Practice.* Swansea: Pineridge Press.

Huh, J. (1999). Dynamic reliability analysis for nonlinear structures using stochastic finite element method. PhD thesis. The University of Arizona.

Huh, J. and Haldar, A. (2002). Uncertainty in Seismic Analysis and Design. *Journal of Structural Engineering, Special issue on Advances in Engineering of Structures to Mitigate Damage due to Earthquakes* 29(1): 1–7.

Huh, J. and Haldar, A. (2011). A novel risk assessment for complex structural systems. *IEEE Transactions on Reliability* 60 (1): 210–218. https://doi.org/10.1109/TR.2010.2104191.

IBC (2000, 2012, 2018). *International Building Code.* International Code Council, Washington DC, United States.

ISO-19900 (2013). *Petroleum and Natural Gas Industries – General Requirements for Offshore Structures.* ISO.

ISO-19901-2 (2017). *Petroleum and Natural Gas Industries – Specific Requirements for Offshore Structures – Part 2: Seismic Design Procedures and Criteria.* ISO.

ISO-19902 (2007). *Petroleum and Natural Gas Industries – Fixed Steel Offshore Structures.* ISO.

Jayaram, N., Lin, T., and Baker, J.W. (2011). A computationally efficient ground-motion selection algorithm for matching a target response spectrum mean and variance. *Earthquake Spectra* 27 (3): 797–815. https://doi.org/10.1193/1.3608002.

Joint Committee on Structural Safety (JCSS). (2006). *Probabilistic Model Code.* https://www.jcss-lc.org/publications/jcsspmc/part_ii.pdf.

Jones, S.W., Kirby, P.A., and Nethercot, D.A. (1980). Effect of semi-rigid connections on steel column strength. *Journal of Constructional Steel Research* 1 (1): 38–46. https://doi.org/10.1016/0143-974X(80)90007-3.

Journée, J.M.J. and Massie, W.W. (2001). Offshore hydromechanics. Delft University of Technology.

Kang, S.-C., Koh, H.-M., and Choo, J.F. (2010). An efficient response surface method using moving least squares approximation for structural reliability analysis. *Probabilistic Engineering Mechanics* 25 (4): 365–371.

Khaloo, A.R., Javanmardi, M.R., and Azizsoltani, H. (2014). Numerical characterization of anisotropic damage evolution in iron based materials. *Scientia Iranica. Transaction A Civil Engineering* 21 (1): 53.

Khuri, A.I. and Cornell, J.A. (1996). *Response Surfaces Designs and Analyses*. New York: Marcel Dekker, Inc.

Kim, S.H. and Na, S.W. (1997). Response surface method using vector projected sampling points. *Structuralc Safety* 19 (1): 3–19. https://doi.org/10.1016/S0167-4730 (96)00037-9.

Kiureghian, D. and Liu (1985). *Structural reliability under incomplete probability information, Research Report No. UCB/SESM-85/01*. Berkeley: University of California.

Kondoh, K. and Atluri, S.N. (1987). Large-deformation, elasto-plastic analysis of frames under nonconservative loading, using explicitly derived tangent stiffnesses based on assumed stresses. *Computational Mechanics* 2 (1): 1–25. https://doi.org/10.1007/ BF00282040.

Krige, D.G. (1951). A statistical approach to some basic mine valuation problems on the Witwatersrand. *Journal of the Southern African Institute of Mining and Metallurgy* 52 (6): 119–139. https://hdl.handle.net/10520/AJA0038223X_4792.

Laplace, P.S.M. (1951). *A Philosophical Essay on Probabilities* (translated from the sixth French edition) (ed. F.W. Truscott and F.L. Emory). New York: Dover Publications.

LATBSDC (2011). *An Alternative Procedure for Seismic Analysis and Design of Tall Buildings Located in the Los Angeles Region*. Los Angeles, CA: Los Angeles Tall Buildings Structural Design Council (LATBSDC).

Lee, S.Y. and Haldar, A. (2003a). Reliability of frame and shear wall structural systems. II: dynamic loading. *Journal of Structural Engineering* 129 (2): 233–240. https://doi.org/10.1061/(ASCE)0733-9445(2003)129:2(233).

Lee, S.Y. and Haldar, A. (2003b). Reliability of frame and shear wall structural systems. I: static loading. *Journal of Structural Engineering* 129 (2): 224–232. https://doi.org/10.1061/(ASCE)0733-9445(2003)129:2(224).

Leger, P. and Dussault, S. (1992). Seismic-energy dissipation in MDOF structures. *Journal of Structural Engineering* 118 (5): 1251–1269. https://doi.org/10.1061/ (ASCE)0733-9445(1992)118:5(1251).

Li, J., Wang, H., & Kim, N. H. (2012). Doubly weighted moving least squares and its application to structural reliability analysis, *Structural and Multidisciplinary Optimization*, 46(1), 69–82, https://doi.org/10.1007/s00158-011-0748-2.

Lichtenstern, A. (2013). *Kriging Methods in Spatial Statistics*. Technische Universität München.

Lin, Y.K. (1967). *Probabilistic Theory of Structural Dynamics*. New York: McGraw-Hill.

Lin, Y.K. and Cai, G.Q. (2004). *Probabilistic Structural Dynamics: Advanced Theory and Applications*. Mcgraw-hill Professional Publishing.

Lionberger, S.R. and Weaver, W. Jr. (1969). Dynamic response of frames with nonrigid connections. *Journal of the Engineering Mechanics Division* 95 (1): 95–114. https://doi.org/10.1061/JMCEA3.0001087.

Loth, C. and Baker, J.W. (2015). Rational design spectra for structural reliability assessment using the response spectrum method. *Earthquake Spectra* 31 (4): 2007–2026. https://doi.org/10.1193/041314EQS053M.

Mai, P.M., Imperatori, W., and Olsen, K.B. (2010). Hybrid broadband ground-motion simulations: combining long-period deterministic synthetics with high-frequency multiple S-to-S backscattering. *Bulletin of the Seismological Society of America* 100 (5A): 2124–2142. https://doi.org/10.1785/0120080194.

Mathisen, J. and Bitner-Gregersen, E. (1990). Joint distributions for significant wave height and wave zero-up-crossing period. *Applied Ocean Research* 12 (2): 93–103. https://doi.org/10.1016/S0141-1187(05)80033-1.

Mehrabian, A. and Haldar, A. (2005). Some lessons learned from post-earthquake damage survey of structures in Bam, Iran Earthquake of 2003. *Structural Survey* 23 (3): 180–192.

Mehrabian, A. and Haldar, A. (2007). Mathematical modelling of a "Post-Northridge" steel connection. *International Journal of Modelling, Identification and Control* 2 (3): 195–207.

Mehrabian, A., Haldar, A., and Reyes-Salazar, A. (2005). Seismic response analysis of steel frames with post-North ridge connection. *Steel and Composite Structures* 5 (4): 271–287. https://doi.org/10.12989/scs.2005.5.4.271.

Mehrabian, A., Haldar, A., and Reyes, A.S. (2005). Seismic response analysis of steel frames with post-Northridge connection. *Steel and Composite Structures* 5 (4): 271–287.

Mehrabian, A., Ali, T., and Haldar, A. (2008). Nonlinear analysis of a steel frame. *Journal of Nonlinear Analysis Series A: Theory, Methods & Applications* 71 (12): 616–623.

Mehrabian, A., Ali, T., and Haldar, A. (2009). Nonlinear analysis of a steel frame. *Nonlinear Analysis: Theory, Methods & Applications* 71 (12): e616–e623. https://doi.org/10.1016/j.na.2008.11.092.

Mirzadeh, J. (2015). Reliability of dynamically sensitive offshore platforms exposed to extreme waves. PhD thesis. University of Western Australia.

Mirzadeh, J., Kimiaei, M., and Cassidy, M.J. (2016). Performance of an example jack-up platform under directional random ocean waves. *Applied Ocean Research* 54: 87–100. https://doi.org/10.1016/j.apor.2015.10.002.

Motamedi, M.H. and Foster, C.D. (2015). An improved implicit numerical integration of a non-associated, three-invariant cap plasticity model with mixed isotropic–kinematic hardening for geomaterials. *International Journal for Numerical and Analytical Methods in Geomechanics* 39 (17): 1853–1883. https://doi.org/10.1002/nag.2372.

Motazedian, D. and Atkinson, G.M. (2005). Stochastic finite-fault modeling based on a dynamic corner frequency. *Bulletin of the Seismological Society of America* 95 (3): 995–1010. https://doi.org/10.1785/0120030207.

Nordenström, N. (1973). A method to predict long-term distributions of waves and wave-induced motions and loads on ships and other floating structures. Thesis (doctoral). Chalmers University of Technology.

Nowak, A.S. and Collins, K.R. (2012). *Reliability of Structures*. Boca Raton, FL: CRC Press.

Perzyna, P. (1966). Fundamental problems in viscoplasticity. *Advances in Applied Mechanics* 9: 243–377. https://doi.org/10.1016/S0065-2156(08)70009-7.

Popov, E.P., Yang, T.S., and Chang, S.P. (1998). Design of steel MRF connections before and after 1994 Northridge earthquake. *Engineering Structures* 20 (12): 1030–1038. https://doi.org/10.1016/S0141-0296(97)00200-9.

Rackwitz, R. (1976). Practical probabilistic approach to design, *Bulletin* No. 112, Comité European du Be'ton (CEB), Paris, France.

Rackwitz, R., and Fiessler, B. (1976). Note on discrete safety checking when using non-normal stochastic models for basic variables. Numerical methods for probabilistic dimensioning and safety calculation. Load Project Working Session, MIT, Cambridge.

Rackwitz, R. and Fiessler, B. (1978). Structural reliability under combined random load sequences. *Computers & Structures* 9 (5): 489–494. https://doi.org/10.1016/0045-7949 (78)90046-9.

Rajashekhar, M.R. and Ellingwood, B.R. (1993). A new look at the response surface approach for reliability analysis. *Structural Safety* 12 (3): 205–220. https://doi.org/10.1016/0167-4730(93)90003-J.

Reyes-Salazar, A. and Haldar, A. (2000). Dissipation of energy in steel frames with PR connections. *Structural Engineering and Mechanics* 9 (3): 241–256. https://doi.org/10.12989/sem.2000.9.3.241.

Reyes-Salazar, A. and Haldar, A. (2001). Energy dissipation at PR frames under seismic loading. *Journal of Structural Engineering* 127 (5): 588–592. https://doi.org/10.1061/(ASCE)0733-9445(2001)127:5(588).

Richard, R.M., Allen, C.J., and Partridge, J.E. (1997). Proprietary slotted beam connection designs. *Modern Steel Construction* 37: 28–33.

Sane, S.M. (2007). Disturbed state concept based constitutive modeling for reliability analysis of lead free solders in electronic packaging and for prediction of glacial motion. PhD thesis. The University of Arizona.

SCEC (2016). *Broadband Platform*. Southern Califorinia Earthquake Center (SCEC).

Schmedes, J., Archuleta, R.J., and Lavallée, D. (2010). Correlation of earthquake source parameters inferred from dynamic rupture simulations. *Journal of Geophysical Research: Solid Earth* 115 (B3): https://doi.org/10.1029/2009JB006689.

Sen, N., Azizsoltani, H., and Haldar, A. (2016). Issues in generating response surfaces for reliability analysis of large complex dynamic systems, EMI/PMC 2016, ASCE, Vanderbilt University, May 22–25, Paper No. 185.

Shariff, A.A. and Hafezi, M.H. (2012). Modelling significant wave height data of North Sea: Rayleigh vs Weibull distribution. *Applied Mechanics and Materials* 157–158: 652–657. https://doi.org/10.4028/www.scientific.net/amm.157-158.652.

Sharifian, H., Bargi, K., and Zarrin, M. (2015). Ultimate strength of fixed offshore platforms subjected to near-fault earthquake ground vibration. *Shock and Vibration* https://doi.org/10.1155/2015/841870.

Shi, G. and Atluri, S.N. (1988). Elasto-plastic large deformation analysis of space-frames: a plastic-hinge and stress-based explicit derivation of tangent stiffnesses. *International Journal for Numerical Methods in Engineering* 26 (3): 589–615. https://doi.org/10.1002/nme.1620260306.

Shields, M.D. (2014). Simulation of spatially correlated nonstationary response spectrum–compatible ground motion time histories. *Journal of Engineering Mechanics* 141 (6): 04014161. https://doi.org/10.1061/(ASCE)EM.1943-7889.0000884.

Sigurdsson, G. and Cramer, E. (1996). *Guideline for Offshore Structural Reliability Analysis-Examples for Jacket Platforms*. Det Norske Veritas, Report.

Skallerud, B. and Amdahl, J. (2002). *Nonlinear Analysis of Offshore Structures*, 323. Baldock: Research Studies Press.

Snee, R.D. (1973). Some aspects of nonorthogonal data analysis: part I. Developing prediction equations. *Journal of Quality Technology* 5 (2): 67–79. https://doi.org/10.1080/00224065.1973.11980577.

Soares, C.G. (1998). Risk and reliability in marine technology. *Oceanographic Literature Review* 7 (45): 1240.

Somerville, P.G. (1997). *Development of Ground Motion Time Histories for Phase 2 of the FEMA/SAC Steel Project*. Washington, DC: SAC Joint Venture Federal Emergency Management Agency (FEMA).

Suárez, L.E. and Montejo, L.A. (2005). Generation of artificial earthquakes via the wavelet transform. *International Journal of Solids and Structures* 42 (21-22): 5905–5919. https://doi.org/10.1016/j.ijsolstr.2005.03.025.

Sudret, B. (2012). Meta-models for structural reliability and uncertainty quantification. *Asian-Pacific Symposium on Structural Reliability and its Applications* (pp. 1–24). https://hal.inria.fr/hal-00683179/.

Taflanidis, A.A. and Cheung, S.H. (2012). Stochastic sampling using moving least squares response surface approximations. *Probabilistic Engineering Mechanics* 28: 216–224. https://doi.org/10.1016/j.probengmech.2011.07.003.

TBI (2010). *Guidelines for Performance-Based Seismic Design of Tall Buildings*. Pacific Earthquake Engineering Research Center, Tall Buildings Initiative (TBI).

Tromans, P.S., Anaturk, A.R., and Hagemeijer, P. (1991). A new model for the kinematics of large ocean waves-application as a design wave. *The First International Offshore and Polar Engineering Conference*, OnePetro, Edinburgh, UK. The International Society of Offshore and Polar Engineers.

Tucker, J.P., Chan, D.K., Subbarayan, G., and Handwerker, C.A. (2014). Maximum entropy fracture model and its use for predicting cyclic hysteresis in Sn3. 8Ag0. 7Cu and Sn3. 0Ag0. 5 solder alloys. *Microelectronics Reliability* 54 (11): 2513–2522. 10.1016/j.microrel.2014.04.012.

Tummala, E.R.R. and Rymaszewski, E.J. (1997). *Microelectronics Packaging Handbook*. Springer.

Uang, C.M., Yu, Q.S., Sadre, A. et al. (1995). Performance of a 13-Story steel moment-resisting frame damaged in the 1994 Northridge earthquake. *Technical Report SAC 95-04*. SAC Joint Venture.

USGS (2008). *Interactive Deaggregations (Beta)*. United States Geological Survey (USGS).

Vazirizade, S.M., Haldar, A., and Gaxiola-Camacho, J.R. (2019). Uncertainty quantification of sea waves – an improved approach. *Oceanography & Fisheries* 9 (5): 10.19080/OFOAJ.2019.09.555775.

Veritas, D.N. (2010). *Environmental Conditions and Environmental Loads. Recommended Practice DNV-RP-C205*. https://home.hvl.no/ansatte/tct/FTP/H2021%20Marinteknisk%20Analyse/Regelverk%20og%20standarder/DnV_documents/RP-C205.pdf.

Vianco, P.T., Rejent, J.A., and Kilgo, A.C. (2003). Time-independent mechanical and physical properties of the ternary 95.5 Sn-3.9 Ag-0.6 Cu solder. *Journal of Electronic Materials* 32 (3): 142–151. https://doi.org/10.1007/s11664-003-0185-0.

Wackernagel, H. (2013). *Multivariate Geostatistics: An Introduction with Applications*. Springer Science & Business Media.

Walling, M., Silva, W., and Abrahamson, N. (2008). Nonlinear site amplification factors for constraining the NGA models. *Earthquake Spectra* 24 (1): 243–255. https://doi.org/10.1193/1.2934350.

Watson-Lamprey, J. and Abrahamson, N. (2006). Selection of ground motion time series and limits on scaling. *Soil Dynamics and Earthquake Engineering* 26 (5): 477–482. https://doi.org/10.1016/j.soildyn.2005.07.001.

Webster, R. and Oliver, M.A. (2007). *Geostatistics for Environmental Scientists*. Wiley.

Wells, D.L. and Coppersmith, K.J. (1994). New empirical relationships among magnitude, rupture length, rupture width, rupture area, and surface displacement. *Bulletin of the Seismological Society of America* 84 (4): 974–1002. https://doi.org/10.1785/BSSA0840040974.

Wheeler, J.D. (1970). Method for calculating forces produced by irregular waves. *Journal of Petroleum Technology* 22 (03): 359–367. https://doi.org/10.2118/2712-PA.

Whitenack, R. (2004). Design and analysis of solder connections using accelerated approximate procedure with disturbed state concept. PhD thesis. The University of Arizona.

Wu and Wirsching (1987). New algorithm for structural reliability estimation. *Journal of Engineering Mechanics, ASCE* 113 (9): 1319–1336.

Xiao, Q. and Armstrong, W.D. (2005). Tensile creep and microstructural characterization of bulk Sn3. 9Ag0. 6Cu lead-free solder. *Journal of Electronic Materials* 34 (2): 196–211. https://doi.org/10.1007/s11664-005-0233-z.

Yamamoto, Y. and Baker, J.W. (2013). Stochastic model for earthquake ground motion using wavelet packets. *Bulletin of the Seismological Society of America* 103 (6): 3044–3056. https://doi.org/10.1785/0120120312.

Yao, T.J. and Wen, Y.K. (1996). Response surface method for time-variant reliability analysis. *Journal of Structural Engineering* 122 (2): 193–201. https://doi.org/10.1061/(ASCE)0733-9445(1996)122:2(193).

Zeng, Q.L., Wang, Z.G., Xian, A.P., and Shang, J.K. (2005). Cyclic softening of the Sn-3.8 Ag-0.7 Cu lead-free solder alloy with equiaxed grain structure. *Journal of Electronic Materials* 34 (1): 62–67. https://doi.org/10.1007/s11664-005-0181-7.

Zimmerman, R.B., Baker, J.W., Hooper, J.D. et al. (2015). Response history analysis for the design of new buildings in the NEHRP provisions and ASCE/SEI 7 standard: part III-Example applications illustrating the recommended methodology. *Earthquake Spectra* 33 (2): 419–447. https://doi.org/10.1193/061814EQS087M.

Zwick, J.W. and Desai, C.S. (1999). Structural reliability of PBGA solder joints with the disturbed state concept. *Pacific Rim/ASME International Intersociety Electronic and Photonic Packaging Conference, Advances in Electronic Packaging (Interpack'99)*, Maui, HI, 13–19 June (pp. 1865–1874).

Index

a

abnormal-level earthquake (ALE)   186
accelerated testing   7, 248
acceleration time history   73, 75, 86
acceptable risk   11
accuracy of simulations   43
added mass   94
advanced factorial design
    (AFD)   116, 121
advanced first-order reliability method
    (AFOSM)   26
allowable deflection for JTP   180, 189
allowable response   47
allowable value   47
alternative to the classical MCS   248
American Institute of Steel Construction
    (AISC)   23, 26
American Petroleum Institute (API)   95
American Society of Civil Engineers
    (ASCE)   210
American Society of Mechanical
    Engineering (ASME)   247
analysis of variance (ANOVA)   56
AND combination   12
anisotropic damage plastic constitutive
    models   250
application boundaries of REDSET   248

Applied Technology Council (ATC)   205
arc length control   200
artificial intelligence   42
assumed stress-based FEM   3, 173, 195,
    197, 213
asymptotic EVDs   21
attenuated with the depth   183
axial point   59, 119, 121
axioms of probability   13

b

ball grid array (BGA)   251
basic factorial designs   58
Bayesian method   5
beam-column connections   204
beam-column elements   67, 176,
    197, 202
beta distribution   19
binomial coefficient   20
binomial distribution   14, 19
bivariate approximation   6
bounded linear model   129
Box and Whisker plot
    182, 190
Box–Behnken designs   55
broadband platform (BBP)   75, 84,
    193, 225

Reliability Evaluation of Dynamic Systems Excited in Time Domain: Alternative to Random Vibration and Simulation, First Edition. Achintya Haldar, Hamoon Azizsoltani, J. Ramon Gaxiola-Camacho, Sayyed Mohsen Vazirizade, and Jungwon Huh.
© 2023 John Wiley & Sons, Inc. Published 2023 by John Wiley & Sons, Inc.

c

California universities for research in earthquake engineering (CUREE) 207

Canadian Standard Association (CSA) 26

capacity reduction factor 23

cap plasticity models 250

center point 51, 54, 59, 113, 114, 121

central composite design (CCD) 55, 112

chain rule of differentiation 49

checking point 27, 113, 114

chi-square test 14

Cholesky factorization 35

classical design 55

classical perturbation method 48

classical plasticity models 252
  Drucker-Prager 252
  Mohr-Coulomb 252
  Tresca 252
  von Misses 252

coded variable space 50

coefficient of determination 57, 129, 133

coefficient of variation 14

coefficients of the thermal expansion 249, 255

collapse mechanisms 44

collapse prevention 66, 74, 166

Comité European du Be'ton (CEB) 26

complex nonlinear dynamic engineering systems (CNDES) 2, 13, 47, 73, 111

computational platform 196

computer chips 7

conditional PDF 18, 105

conditional probability 13

confidence interval 262

connections 200
  Type I 200
  Type II 200
  Type III 200

consequences of failure 192

constitutive modeling calibration 255

constitutive modeling techniques 250

constitutive relationship 204

constrained new wave (CNW) concept 100, 175

continuous RVs 19
  beta distribution 19
  lognormal distribution 19
  normal distribution 19

continuous yielding behavior 250

correlated non-normal variables 37

correlated random variables 28, 33

correlation coefficient 18

covariance function 127

covariogram 127

creep deformation 251

creep models 254
  elasto-viscoplastic (evp) 254
  viscoelastic (ve) 254
  viscoelastic-viscoplastic (vevp) 254

creep-related model parameters 256

critical disturbance (Df) set 260

critical state continuous yielding models 250

cumulative distribution function (CDF) 14

cyclic thermomechanical loading 249

d

damage-tolerant structures 8

damping matrix 197

decomposition methods 6

degree of polynomial 51, 52

Delta stretching 98

design point 27, 48

design seismic loading 185

design spectrum 161

design time histories 185

design wave loading 185

deterministic design 12

deviatoric plastic strain 254

different design alternatives   183
different environments   183
directionality   179
directionality of wave loading   193
discrete RVs   19
  binomial distribution   19
  geometric distribution   19
  Poisson distribution   19
displacement-based FEM   3, 195,
    196, 213
dissimilarity function   129, 131
dissipate excessive heat   249
distributed state concept (DSC)   250, 253
distributions   14
  binomial   14
  lognormal   14
  normal   14
  Poisson   14
disturbance function   254
disturbance-related model
    parameters   256
drag coefficient   94, 179
dwell time   256
dynamic amplification of responses   238
dynamic degrees of freedom (DDOFs)   4
dynamic governing equation of
    motion   197
dynamic lateral loading   183
dynamic loading   203
  loading   203
  reloading   203
  unloading   203
dynamic plastic yield function   255
dynamic properties   183
dynamic responses   183
dynamic tangent stiffness matrix   198

e

earthquake time history   73
edge point   59, 119
efficiency of simulations   43

efficient MCS method   5
elastic-perfectly plastic   204
elastoplastic material models   254
electronic packaging (EP)   73, 247
element level   4, 71
energy-related issues   185
environmental damages   185
epicenter   88
equiradial designs   55
equivalent lateral load procedure   72
equivalent normal transformation
    three-parameter   33
    two-parameter   31
experimental design   55
experimental region   54
experimental sampling design
    schemes   52, 54
experimental sampling points   55
experimental variogram   120, 129, 131
  bounded linear model   129
  exponential model   129
  nugget effect model   129
  spherical model   129
explicit design space
    decomposition (EDSD)   6
explicit limit state function   47, 111
exponential distribution   19
exponential model   129
exposure level   186
external load vector   197
extreme-level earthquake (ELE)   186
extreme value distributions   20, 21
  Type I   21
  Type II   21
  Type III   21

f

factorial (factored factorial) design   55
factorial point   121
failure mode approach   40
failure surface   24

fault tree diagram 40
Federal Emergency Management Agency
    (FEMA) 6, 140, 207
finite difference method 48
finite element method (FEM) 2
first-order reliability method (FORM) 8,
    21, 25, 35
first-order second-moment method
    (FOSM) 25
first-order Taylor series expansion 4
first quantile 182
flexible connections 195
flexible post-Northridge
    connections 195
fluid damping 94
Forristall wave height distribution 106
FORTRAN language 4
Fourier spectrum 86
four-node quadrilateral plane stress
    elements 251
frame-type structure 3, 6, 197
frequency contents 182, 183
frequency diagram 14, 16
full distributional approach 25
fully adjusted (FA) state 254
fully restrained 6, 200
fundamental period 79, 187

g
generalized least squares technique 122
geometric distribution 19
geostatistical method 126
gravity loads 71
    dead load 71
    live load 71

h
hardening function 253
Hasofer-Lind reliability index 28, 31
heating and cooling of computers 247
hierarchical single-surface (HISS) 250

high-dimensional model
    representation (HDMR) 6
high-performance computing 181, 189
histogram 14
hybrid broadband simulation 84
hydrodynamic coefficients 179
hydrodynamic damping 94, 179
hydrostatic stress 254
hyperbinomial distribution 20
hypocenter 84, 88, 215, 238

i
idle time 256
immediate occupancy 66, 166
implicit limit state function 111
implicit RS 48
improved response surface (IRS) 51, 111
inertia coefficient 94, 179
innovative deterministic approaches 8
intelligent simulations 8
interaction equations 67
internal force vector 197
International Building Code (IBC) 210
international standard organization
    (ISO) 185
intersection of events 13
inter-story drift 66, 204, 212
iterative perturbation method 48

j
jacket-type offshore platform
    (JTOP) 92, 175
joint CDF 17
joint distribution 18
joint distribution of multiple RVs 18
joint PDF 17
Jonswap spectrum 97

k
Karhunen-Loeve Orthogonal
    Expansion 4
Kolmogorov-Smirnov test 14, 16

Kriging method   9, 126
  ordinary   127
  simple   127
  universal   127

l

laboratory testing   7, 248
Lagrange multiplier   128, 130
lateral deflection   66, 204
lead-free solders   247, 248
least squares method   122, 131
life cycle testing   249
life expectancy of solder balls   264
life safety   66, 74, 166, 195, 207
life safety category   186
limit state equation   24
limit state function   2, 26, 47
linear unbiased surrogate   127
load and resistance factor design   22, 151
load factors   23
lognormal distribution   14, 19
loss of integrity or collapse   186

m

magnification factor MFT   256
magnification factor, MFTD   256
major paradigm shift   210
marginal PDF   18
marine growth   179
Masing rule   203
mass matrix   197
  consistent   197
  lumped   197
material nonlinearity   7
mathematics of probability   14
maximum considered earthquake
    (MCER)   74
maximum likelihood   131
maximum response information   47
mean   14
mean sea water level   176

mean value first-order second-moment
    (MVFOSM) method   25
mechanically equivalent
    formulations   26
median   182
method of maximum likelihood   16, 131
method of moments   16
microcracks   249
microprocessor   249
minimum norm quadratic   131
modal approach   72
model reduction techniques   8
modified advanced factorial design
    scheme   119
modified external load vector   199
modified Newton-Raphson method   200
modulus of elasticity   91
monotonic loading   201
Monte Carlo simulation   5, 8, 41, 42,
    48, 248
Morison equation   95
most probable failure point (MPFP)   28,
    48, 113
most significant random variables   181
moving least squares method   122, 123
M–θ curves   200
  bi-linear   201
  cubic B-spline model   201
  exponential model   201
  linear   200
  piecewise linear model   201
  polynomial model   201
  Richard model   201
multicomponent disturbed state concept
    (MDSC)   254
multiple deterministic analyses   12
multiple RVs   17

n

National earthquake hazard reduction
    program (NEHRP)   74, 140, 239

natural frequencies   184
Neumann expansion   4
Newmark Beta method   198
Newmark's step-by-step direct
    integration   197
Newton–Raphson recursive
    algorithm   35, 40, 48
new wave (NW) theory   96
nodal thermal displacements   251
non-homogeneity   238
nonlinear dynamic FE analyses
    (NDFEA)   114
nonlinearities   197
  geometric   197
  joint conditions   197
  material   197
nonlinear site effects   84
nonlinear structural analysis   49
nonlinear time domain analysis   72
nonlinear viscoplastic deformations   250
normal distribution   14, 19
Northridge earthquake   139
nugget effect model   129
numerical experimentations using
    simulation   42

o

offshore structures (OFS)   91, 175
onshore structures (ONS)   91, 175
optimum number of TNDFEA   122
OR combination   12
ordinary   127
ordinary kriging   127
origin shift function   252
overall lateral deck deflection   189, 212
overall lateral displacement   66

p

Pacific Earthquake Engineering Research
    Center (PEER)   75, 166, 193, 225
panel zone deformation   201

parallel processing technique   49,
    181, 262
parallel system   44
partially restrained   6, 200
partially restrained connections   213
  post-Northridge type   213
  pre-Northridge type   213
peak ground acceleration (PGA)   73, 215
performance-based seismic design
    (PBSD)   6, 66, 135, 166, 195, 207
performance function   24, 47
performance levels   207
  collapse prevention (CP)   66, 166, 208
  immediate occupancy (IO)   66,
    166, 208
  life safety (LS)   66, 166, 208
permissible response   47
permissible value   47
phase change   250
physical mechanical support   249
physics-based behavior   250, 255
physics-based modeling   2, 3, 206, 244
plasticity models   252
  Drucker-Prager   252
  Mohr-Coulomb   252
  Tresca   252
  von Misses   252
plastic strain   253
Poisson distribution   14, 19
polynomial chaos   4
poor soil conditions   238
post-failure behavior   67
post-Northridge connections   205, 206
  bottom haunch   205
  connections with vertical ribs   205
  cover-plated   205
  dog-boned   205
  reduced beam sections (RBS)   205
  sided-plated   205
  slotted-web beam-column
    connections   205

slotted-web moment connection 205
spliced beam 205
post-Northridge design
    requirements 139
power spectral density function 4
pre-Northridge connection 205, 206
pre-Northridge design guidelines 158
principal stress space 250
probabilistic sensitivity indices 39
probability density function (PDF) 14
probability mass function (PMF) 14
probability of exceedance 74, 210
probability of failure 23, 139, 147
probability theory 12
progressive failure 67

q
qualitative measure 23

r
ramp time 256
random variables 14
    continuous 14
    discrete 14
random vibration 4, 8
range parameter 132
Rayleigh distribution 105
Rayleigh-type damping 197
reduced number of RVs 115
reduced order model 5
regression analysis 9, 55, 121
regression coefficients 52
regressor variable 56
relatively intact (RI) state 254
reliability index 25, 27, 39
resonance condition 185, 193, 239
response surface (RS) 6, 47
response surface method (RSM) 49, 111
response variable 56
return period 75, 210, 215
Richard four-parameter-model 201

curve shape parameter (N) 201
initial stiffness (k) 201
plastic stiffness (k_P) 201
reference moment (M_0) 201
risk coefficients, C_R 75
risk-targeted maximum considered
    earthquake 74, 166
rupture generation concepts 84

s
SAC 140, 147, 207
safety index 25, 27
sampling design schemes 48, 51, 113
    MS2, 119
    MS3, 121
    MS3-KM 133, 180
    MS3-MLSM 124
    S1, 113, 116
    S2, 116
    S3, 117
    S3-KM 133
    S3-MLSM 124
sampling points 55
sampling scheme 48
saturated design (SD) 55
scale factor (SF) 75, 78
sea water level 183
sea wave states 180
second-order polynomial
    with cross terms 53
    without cross terms 53
second-order reliability method
    (SORM) 25
second-order Taylor series expansion 4
seismic loading 8
seismic load-tolerant structures 237
seismic reserved capacity factor 186
seismic risk category (SRC) 186
Seismic Safety Council (BSSC) 210
Seismic Structural Design Associates
    (SSDA) 205

seismic zone   186
sensitivity-based analysis   48
   classical perturbation   48
   finite difference   48
   iterative perturbation method   48
sensitivity index   39, 115, 119
serviceability LSF   66, 180, 189
set theory   12
   collectively exhaustive events   13
   complementary event   12
   event   12
   intersection of events   12
   mutually exclusive events   13
   null set   12
   sample points   12
   sample space   12
   union of events   12
severity of an earthquake   196
SFEM-dynamic   3
SFEM-static   3, 4
shape functions   3
shear wave velocity   76, 84
significant wave height   96, 181
silicon chip   250
sill parameter   132
simple kriging   127
simulation region   145
single-plane fault surface   85
site class   161
site classes A–F   239
site coefficients   186
skewness   14
skewness coefficient   14
Sn-3.9Ag-0.6Cu (SAC) lead-free
   solders   249
soil class   76
soil classification   76
soil condition   196
soil-structure systems   238
solder balls   7, 73

sources of nonlinearity   185, 204
   connection conditions   204
   large deformations   204
   material properties   204
Southern California Earthquake Center
   (SCEC)   75
space reduction technique   5, 44, 48
spectral acceleration   76, 186
spherical model   129
splash zone   176, 182, 185, 193
stable configuration approach   40
standard beta distribution   20
standard deviation   14
star points   56
statistical characteristics   187
statistical correlation   67
statistical information   17, 112, 113. *see
   also* uncertainty quantification
statistically independent   13, 18
statistical tests   14
stochastic dimension reduction
   technique   5
stochastic finite element
   method (SFEM)   2
stochastic phases   5
strain-hardening ratio   91, 179
strain rate   250
strength formulation   26
strength LSF   66, 180, 181
stress-based FEM   4
stress formulation   26
stress ratio function   253
structural damping   94
structural dynamic properties   204
   damping   204
   frequency   204
   mode shape   204
Structural Engineering Association of
   California (SEAOC)   205, 207
submerged condition   183

submerged fundamental period   187
substrate   250
suitability factor   79, 187
sum of the squares errors (SRSSE)
      79, 91
supplementary variables   131
support vector machines (SVM)   6, 7
surface soil   186
surrogate model   5, 126
S-variate approximation   6
system level   4, 71, 115
system reliability   40, 67

t

tangent stiffness matrix   3, 197
target probabilistic ground motion
      response spectrum (TPGRS)   74, 75,
      161, 187
Taylor series expansion   4, 25, 52
temperature dependent modeling   255
thermal cyclic fatigue   249
thermal displacement   256
thermomechanical loading   7, 8, 73,
      247, 256
third quantile   182
three-dimensional plot   17
three-directional (3D) behavior   193
Tin-Lead (Sn-Pb) solder   249
Tin-Silver-Copper solder   249
total lateral drift   180, 181
total nonlinear dynamic FE analyses
      (TNDFEA)   115, 117, 181, 189,
      190, 263
total sum of squares   57
trapezoidal distribution   16, 21
triangular distribution   16, 21

u

ultimate strength design   22
uncertainty management   193
uncertainty propagation   39, 115

uncertainty quantification   14
   coefficient of variation (COV)   14
   mean   14
   skewness   14
   skewness coefficient   14
   standard deviation   14
   variance   14
Uniform Building Code   74
uniform distribution   16, 21
uniform hazard response spectrum
      (UHRS)   75
Uniform Hazard Spectrum (UHS)   240
uniform shell designs   55
union of events   13
United States Geological Survey
      (USGS)   75, 76
univariate approximation   6
universality conditions   130
universal kriging   127, 133

v

variance   14
variance reduction techniques   44
variogram cloud   129, 131
variogram function   128, 129
Venn diagram   13
viscous damping   184
viscous fluid   72, 91
volumetric plastic strains   250

w

wave height   96
wave loading   8
wave profile   96
weakest link system   44
Weibull distribution   21, 105, 181
weighted average   126
weight factor   123
   constant   123
   exponential   123
   fourth-order polynomial   123

weight factor (*cont'd*)
  linear   123
  quadratic   123
Wheeler stretching   98
wind tunnel testing   72
working stress design   22

y
yield stress   91

z
zero-up crossing period   97